THE SHORTER
SCIENCE AND CIVILISATION IN CHINA

COLIN A. RONAN

The Shorter
Science and Civilisation
in China

AN ABRIDGEMENT OF
JOSEPH NEEDHAM'S ORIGINAL TEXT

Volume 1

VOLUMES I AND II OF THE
MAJOR SERIES

Cambridge University Press

CAMBRIDGE

LONDON   NEW YORK   NEW ROCHELLE
MELBOURNE   SYDNEY

Published by the Press Syndicate of the University of Cambridge
The Pitt Building, Trumpington Street, Cambridge CB2 1RP
32 East 57th Street, New York, NY 10022, USA
296 Beaconsfield Parade, Middle Park, Melbourne 3206, Australia

First published 1978
First paperback edition 1980

Printed in Great Britain at the
University Press, Cambridge

Library of Congress Cataloguing in Publication Data

Needham, Joseph, 1900–
The shorter Science and civilisation in China.

Bibliography: p.
Includes index.
1. China – Civilisation. 2. China – Intellectual life.
3. Science – China – History. I. Ronan, Colin A. II. Title.
DS721.N392   509'.51   77-82513

ISBN 0 521 21821 7 hard covers
ISBN 0 521 29286 7 paperback

CONTENTS

ILLUSTRATIONS

TABLES

PREFACE

There is no doubt that Dr Joseph Needham's 'Science and Civilisation in China' is a monumental piece of scholarship, and one which breaks new ground in presenting to the Western reader a plentifully detailed and coherent account of the development of science and technology in China from the earliest times until the advent of the Jesuits and modern science in the late seventeenth century. It is a vast work that is of necessity more suited to the scholar and research worker than the general reader. Yet so significant are its contents in presenting something hitherto unavailable in the Western world that there is clearly a need for an abridged version written in a way that will make it available to the general reader whether or not he or she has any scientific training. This both Joseph Needham and the Cambridge University Press have realised, and it is my privilege and pleasure to have been invited to do this: privilege because I am thereby a link in the chain of bringing this almost superhuman study to a wider audience, and pleasure since the task, though not a light one, has meant a voyage of discovery for me across previously uncharted oceans.

Many people have helped and encouraged me with this abridgement, but none more than Joseph Needham himself. Consultations with him, often over tea, have always been an education as well as a happy experience, and have helped me to sort out the problems involved in abridging a text so full of information. I am also indebted to Joseph Needham not only for preparing the bibliography but also for going through my entire text with me. We have read it and he has approved it, and although pressure of work on the main text of 'Science and Civilisation in China' has prevented him from systematically introducing new material, between us we have taken the opportunity to add certain fresh facts and make some modifications of emphasis in the light of knowledge gained since the full text of the earlier volumes appeared. We have also made a few corrections here and there. Moreover, we have changed the chapter headings and altered the order of subjects so that these are not quite the same as in Volumes I and II of the original. However, this is not in any way a second edition of 'Science and Civilisation in China'; it cannot incorporate the new findings of the past twenty years in, for instance, the pre-history of the Chinese culture-area, or in archaeology. It is essentially an abridgement of the work as it stands now.

The romanisation here adopted remains that of Wade–Giles with the substitution of an *h* for the aspirate apostrophe.

My thanks are also due to Mr Anthony Parker of the Cambridge University Press for introducing me to the idea of acting as abridger, and my warmest thanks are also due to the Press' editor Dr Alan Winter, without whose patience my text would have had more blemishes than it carries at the moment. My deep thanks are also due to Mrs Sandra Welch and Mrs Diana Brodie who have typed the manuscript impeccably in spite of it being peppered with alterations, and not least to my wife Penny who has read through every line and made helpful comments.

Bar Hill, Cambridge
23 December 1976

Colin A. Ronan

Fig. 1. General map of China.

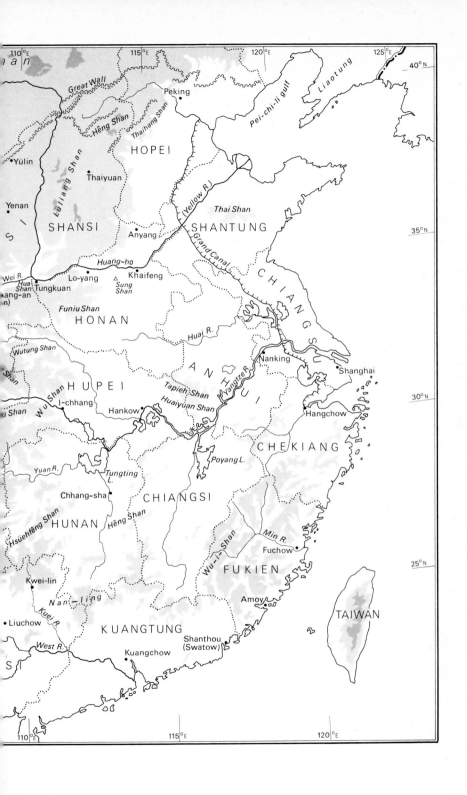

1

Introduction

Today the history of science is becoming recognised as an integral part of the whole history of human civilisation; an essential ingredient in the development of human culture. Tracing its growth has occupied a host of scholars who, almost without exception, have worked backward from the science and technology of today to its cradle in the thought and practice of Mediterranean antiquity. They have uncovered an evolution starting from the Sumerians, Babylonians and Egyptians, an evolution that led to the growth of scientific thought and observation of the natural world among the Greeks and in the Roman Empire. From here they have tracked its transmission to mediaeval Europe by way of Islam, and seen how its arrival led to the revolutionary changes that occurred in the wake of the Renaissance.

To a great extent, all this is something new. A century and a half ago, the scientific contributions of the Sumerians and Babylonians, for instance, were quite unsuspected. In 1837, when William Whewell wrote his memorable *History of the Inductive Sciences*, he could display a bland unconsciousness of any contributions by other civilisations to the scientific culture of the modern West, and do so without criticism. Yet now the situation is somewhat different; not only are the Babylonians and Sumerians recognised, but there is also some appreciation of the legacy we owe to India. All the same, there is still one vast gap in our understanding of our debts to other civilisations – the contribution from Asia, and especially the northernmost of its two oldest civilisations, the Chinese. This scientific heritage is the theme of this book.

Precisely what the Chinese contributed to science and technology as well as to scientific thought depended, as will become clear, on the historical period being considered. In ancient and mediaeval times it was of immense importance, but its character changed after the visit of Jesuit missionaries to Peking early in the seventeenth century, and it gradually fused into the universal science that has continuously developed over the last three hundred years. Before the Jesuit arrival Chinese science was quasi-empirical – based on observation and experience, with theory in a comparatively undeveloped state. Yet, even so, the Chinese succeeded in anticipating many scientific and technical discoveries of the Greeks; they managed to keep pace with the Arabs who had all Greek knowledge at their disposal; and between the first and thirteenth centuries A.D. reached a level of scientific knowledge unapproached in the West.

To those of us brought up in a culture which has the classical world as its foundation, the Chinese achievement may well seem nothing less than astonishing. Certainly there was no rise of modern science in sixteenth-century China as occurred in Europe from this time onwards, while it is also true that the Chinese suffered from a weakness in theoretical ideas and a lack of deductive geometry – the very essence of precision in Greek science. Yet in spite of all this we see in ancient China a society more amenable to the application of science than was the case in Greece, in Rome, or even in mediaeval Europe. What is more, in China there developed an organic philosophy of Nature that closely resembles that which modern science has been obliged to adopt after three centuries of scientific materialism. How this could be under the circumstances is one of the questions that will be discussed.

To appreciate what the Chinese accomplished can be difficult, even today, because unfortunately many misconceptions about Chinese discoveries and scientific development still exist. The old legendary chronology relayed by the seventeenth-century Jesuits lives on, with the result that either too much or too little is ascribed to East Asian origin, and Chinese as well as Western scholars have sometimes been known to ignore, or at least pay scant attention to, achievements made early on in China. Frequently, too, legends themselves are unrecognised as such, with the result that worthwhile evidence is glossed over. Nevertheless, most Europeans are at least dimly aware of a vast and complex civilisation no less intricate and rich than their own, at the other end of the huge Euro-Asian land mass. The main barrier to a more intimate understanding, especially when it comes to Chinese science and technology, has been the Chinese use of ideographic characters. Inevitably most sinologists have been of literary tastes and training and, in consequence, there is a vast amount of scattered literature that has hardly been surveyed, let alone studied in detail.

In one sense, then, this book can be no more than a reconnaissance, and a brief reconnaissance at that, but at least it is based on a detailed approach to some sources made by Joseph Needham and his collaborators. Uniquely qualified as a research scientist and historian of science familiar with the country and its language, and in contact with many

Chinese scientists and scholars, he has been able to study the original Chinese texts from which translations have been made as well as those for which no translations have so far been available. This has enabled him to correct mis-translations and misconceptions. For example, in the only complete translation of the important *Mo Tzu* (Book of Master Mo Ti) from the fourth century B.C., there is a reference to textile manufacture. The accepted translation runs: 'Women work at variegated embroidery: men work at the weaving of stuffs with inserted patterns.' Taken at its face value this looks like a reference to the drawloom; only a careful study of the text shows that nothing is said of inwoven figured patterns. The author is, in fact, referring to 'cut and engraved' work; in other words to a kind of brocade made by stitching coloured threads into an already woven fabric. Thus the drawloom is not concerned and one cannot, therefore, claim evidence for its invention in the fourth century B.C. on this ground. But other evidence does date back some form of it to that time. Plenty of other examples could be quoted but this alone may serve to underline the dangers that face anyone unfamiliar either with the language or with the techniques involved.

Here it should be emphasised that although ninety per cent of the work as published has been written by Joseph Needham, the project, he tells me, would have been absolutely impossible without the partnership of a number of colleagues. From 1948 to 1958 his chief collaborator was Wang Ching-Ning (Wang Ling), now professor at Canberra in Australia, a historian and mathematician; and since 1958 an even older friend, Lu Gwei-Djen, a specialist on the history of medicine and biology. Several other Chinese scholars have also collaborated, notably Ho Ping-Yü of Brisbane, historian of astronomy, alchemy and early chemistry, Lo Jung-Pang in California who contributes, for instance, the chapter on the salt industry and the epic of deep borehole drilling, Chhien Tshun-Hsün at Chicago, one of the best authorities in the world on the history of paper and printing, and Li Li-Shêng, studying the traditional chemical industries. As time has gone on it has been necessary to enlarge more and more the circle, so that textile technology, for example, is now in the charge of Ohta Eizō in Kyoto, and ceramics undertaken by Chhü Chih-Jen in Hong Kong. Western collaborators have also participated in the project, notably Kenneth Robinson, who drafted the physical acoustics section, Derk Bodde, who is surveying the world-outlook of the traditional Chinese literati, and Janusz Chmielewski who is writing the important study of Chinese logic. This is by no means a full list of all collaborators and participators, but it may give some idea of the scope of the team-work that has been involved.

Like the full seven-volume study in perhaps twenty separate parts that

Needham and his collaborators are writing and of which ten parts have now appeared, this abridgement may be, it is hoped, a contribution to international understanding. The genius of the Chinese people has so often been represented to the West as primarily artistic and agricultural: the long succession of technical discoveries taken over from China during the first thirteen centuries of our era has been almost entirely overlooked. How much this genius may have influenced the seventeenth-century scientific revolution in the West has yet to be fully assessed. Even so, one must recognise that all the foundations of our knowledge of electro-magnetics were laid in China, and Europe at the turning-point was greatly affected by the Chinese conviction of the infinity of the universe. Whatever the final answer may be, a knowledge and appreciation of the achievements of the scholars and craftsmen of other cultures can only lead to a growth of mutual comprehension. After all, we must be on our guard against the temptation of thinking that the whole of modern civilisation began with Renaissance figures like Galileo and Vesalius in the sixteenth and seventeenth centuries, and the conclusion that 'Wisdom was born with us'. There was a Chinese contribution to Man's understanding of Nature and his control over it, and it was a great one. No single people or group of peoples has had a monopoly in contributing to the development of science. All achievements should be recognised, and celebrated, if we are to move on our way to a universal brotherhood of Man.

2

The Chinese language

Before describing any of the achievements of Chinese science and technology, it is necessary to have some kind of cultural background against which they can be set. That is the purpose of this volume – to provide a backcloth for the scientific details that the remaining volumes will contain. To form this background it will be best if we glance first at the geography of China, then at its history, and next consider what opportunities there were for an interchange of ideas between Eastern Asia and the West. We shall then be in a position to trace the origin and development of scientific thought in Chinese philosophy – something that is vital if we are to see the inventive genius of China in anything like a proper perspective.

Throughout this volume we shall, of course, have occasion to use Chinese names and refer to Chinese words, so as a preliminary to our background reconnaissance, it is desirable to spend a few moments on the Chinese language itself. To begin with we shall need to be able to transform Chinese words into our own romanised script so that we can write them and, at the same time, get some idea of their pronunciation. Since Chinese is a tonal language where the tone can give different meanings to a single word, any system we adopt is bound to be at best only approximate, and there has been, and still is, controversy about how romanisation should be done. A large number of competing systems grew up, some stemming from a romanisation of the Cantonese dialect once used by the Chinese Post Office, others based on phonetic and linguistic studies. Sir Thomas Wade in 1867 tried to formulate an internationally acceptable system, but French and German sinologists evolved methods based on their own way of pronouncing the Latin alphabet. Nevertheless the Wade–Giles system, so called because it was modified and adopted by H. A. Giles in the 1890s, is that most widely used in the Western world today. Its chief rival perhaps is the Phin-yin system, an alternative introduced officially by the Chinese Government

5

Table 1. *Romanisation of Chinese sounds*

The 24 Consonantal Initials

Wade–Giles system	Phin-yin system	System adopted in this book	Pronunciation
ch-	*zh-* or *j-*	*ch-*	between *chair* and *jar*
ch'-	*ch-* or *q-*	*chh-*	as in *much harm*; strongly aspirated
f-	*f-*	*f-*	as in *farm*
h-	*h-*	*h-*	Gaelic *-ch*, as in *loch*
hs-	*x-*	*hs-*	a slight aspirate preceding and modifying the sibilant, the latter being the stronger. Try dropping the first *i* in *hissing*
j-	*r-*	*j-*	French *j-* as in *je* or *jaune*. A *j-* pronounced at the front of the mouth gives an impression of *r-* (cf. Polish *rz-*)
k-	*g-*	*k-*	between *k* and *g*
k'-	*k-*	*kh-*	*k* strongly aspirated, as in *kick hard*
l-	*l-*	*l-*	as in English
m-	*m-*	*m-*	as in English
n-	*n-*	*n-*	as in English
p-	*b-*	*p-*	like *b* in *lobster*, or Fr. *peu*
p'-	*p-*	*ph-*	as in Irish dialectal pronunciation of *party* or *parliament*, more strongly aspirated than anything in French, German or English
s-	*s-*	*s-*	as in English
sh-	*sh-*	*sh-*	as in English
ss-	*s-*	*ss-*	only occurs with *-ŭ*, q.v.
t-	*d-*	*t-*	nearer *d-* than *t-* in English but not quite *d-*
t'-	*t-*	*th-*	strongly aspirated *t-*, as in Irish dialectal pronunciation of *torment*
ts-	*c-*	*ts-*	as in *jetsam, catsup*
ts'-	*z-*	*tsh-*	*ts-* strongly aspirated, as in *bets hard*
tz-	*c-*	*tz-* ⎫	only occur with *-ŭ*, q.v. Sounds near
tz'-	*z-*	*tzh-* ⎭	to *ts'*
w-	*w-*	*w-*	as in English, but faint
y-	*y-*	*y-*	as in English, but faint

The 42 Vowel, Diphthong and Consonant Finals

-a or *a*	as in *father*, the 'broad' *a*
-ai	as in *aye*, or better Italian *hai, amai*. English *why*
-an	somewhat like Dutch *Arnhem* pronounced by an Englishman, the *r* being unsounded. Or German *ahnung*
-ang	the *-ng* has a partly nasalising and partly gutturalising influence on the vowel. Something like German *angst*
-ao	as in Italian *Aosta, Aorno*. Not so fused as in English *how*

Table 1. *Romanisation of Chinese sounds (continued)*

-ê	nearest approached by English vowel-sound in *earth*, *perch*, or *lurk*
-êi	the foregoing, says Wade, followed enclitically by -*y*. English *money* omitting the -*on*-. Generally sounded as -*ei* or -*ui* (see below)
-ei	generally indistinguishable from English *may*, *play*, *grey*, *whey*
-en	as in English *yet*, *lens*, *ten*
-ên	as in English *bun*
-êng	as in English *unctuous*, *flung*
-erh	as in English *burr*, *purr*
-i	vowel-sound as in English *ease*, *tree*
-ia	not like *yah* but with the vowels more distinct, though not so much so as in Italian *Maria*, *piazza*, and not separately accented
-iai	as in Italian *vecchiaja*
-iang	like -*ang* above, with the additional vowel
-iao	like -*ao* above, with the additional vowel
-ieh	as in French *estropié*
-ien	with vowels distinct, as in Italian *niente*
-ih	short vowel, as in *cheroot*
-in	short vowel, as in English *chin*
-ing	short vowel, as in English *thing*
-io	short vowel, as in French *pioche*
-iu	always longer than English termination -*ew*, e.g. 'chyew' instead of *chew*. *Mew* of cat, as onomatopoeically pronounced
-iung	like -*ung* below, with the additional vowel
-o	something between the vowel-sounds in English *awe*, *paw* and *roll*, *toll*
ong	as in English *dong* cut short
-ou	really -*eo*, English *Joe*
-u	as in English *too*
-ü	as in French *eût*, *tu*
-ŭ	between the *i* in English *bit* and the *u* in *shut*. Only occurs with *ss*-, *tz*- and *tzh*-, 'which it follows from the throat' says Wade 'as if the speaker were guilty of a slight eructation'.
-ua	as in Spanish *Juan*; may contract almost to *wa*
-uai	as in Italian *guai*
-uan	like -*an* above, with the additional vowel
-üan	the *ü* as above, the -*an* as in English *antic*
-uang	like -*ang* above, with the additional vowel
-üeh	as in French *tu es*
-uei	the -*u* as above, the -*ei* as above; cf. French *jouer*
-ui	as in Italian (not French) *lui*
-un	as in Italian *punto*, *lungo*
-ün	as in German *München*
-ung	as in Lancashire dialectal pronunciation of English *bung*, *sung*; not so broad as -*oong*
-uo	the *o* as in English *lone*; the whole as in Italian *fuori*

It should be mentioned that the Wade–Giles system took the sounds of the Peking dialect of 'mandarin' as standard. But the *pu thing hua* pronunciation of today is not quite the same, and we have modified the transcriptions of the present book in a desire to concord with it. Thus we find the circumflex accent essential in words such as *pên* (origin) and *Chêng* (family name), but not in *jen* (a person), *chen* (true) or *Chhen* (family name). Similarly, we retain the diaeresis in *hsü* and *hüan* but we do not write it in *yuan*. The sound represented by the inverted circumflex *ŭ* can always be recognised from the consonants which invariably accompany it; we therefore dispense with this diacritical mark.

in 1962. In the previous table both systems are given, together with the scheme adopted by Joseph Needham and collaborators, which is the one in these volumes. It avoids the apostrophes that Wade and Giles used for aspiration, and replaces them by *h*, thus permitting direct comparison with the sounds of Indian languages where, for instance, 'Buddha' and 'Buddhism' are common romanisations.

As a language Chinese is unique: it is the main one which has remained faithful to the ideographic form of writing. Why it has done so is uncertain; perhaps it refused to move to an alphabetical system because it may originally have been monosyllabic. Whatever the reason, it has retained the ideographic script, unlike what happened, for instance, in ancient Egypt and Sumeria.

The most primitive elements of Chinese are pictographs – drawings reduced to the bare essentials, conventionalised and, in due course, highly stylised. Natural objects, such as celestial bodies, animals, plants, tools and implements, lend themselves most easily to such drawings, and a number of them are shown in Table 2. The ancient forms of characters are often of considerable interest from the standpoint of Chinese science and technology, as will become evident in later volumes. As the language developed, other written characters were adopted. *Indirect symbols* were introduced, being derived by using gestures for actions, effects for causes and so on. Thus, as Table 2 shows, *chih*, to mount, originated from a picture of two footprints pointing upwards, and *fu*, meaning full or blest, is derived from an ancient picture of a jar. There were also *associative compounds*: *fu*, father, consists of ancient signs for hand and stick; *fu*, wife, signs for woman and broom, while the ideograph *nan*, for male or man, comes from the radicals for plough (or strength) and field.

Besides indirect symbols and associative compounds, some sinologists recognise what may be called *mutually interpretative compounds*. Here one sign is derived from another, although originally they both meant the same thing; only later becoming interpreted differently. For instance, *khao*, meaning examination, is said to be derived from *lao*, old, because it is the old who examine the young, yet originally the two characters meant the same.

At the present time there are about 2000 pictographs, indirect symbols, associative compounds and mutually interpretative compounds. Yet these do not exhaust the Chinese characters that may be used. Chinese is also very rich in homophones – words having different meanings but sounding the same (like the English words sew, sow, so). Because of this there was always a tendency to use one ideograph with the sense that properly belonged to another that looked different but sounded the

Table 2. *The development of Chinese script*

PICTOGRAPHS					
Archaic script	Small seal	Modern script	Forms in writing	Meaning	Rad. no.
	人	人	人	*jên,* man	9
		仉 虎	人 虎	*hu,* tiger	141
		羊 象	扁 羊 象	*yang,* sheep	123
		象 魚	象 魚	*hsiang,* elephant	—
		魚 壺	魚 壺	*niao,* bird	196
		壺 車	壺 車	*yü,* fish	195
		月	月	*hu,* wine-vessel	—
		山	山	*chhê,* chariot, car	159
				yüeh, moon	74
				shan, mountain	46

INDIRECT SYMBOLS

射 *shê,* to shoot with a bow; 伐 *fa,* to attack (man being decapitated); 爲 *wei,* lead, manage, do (hand leading an elephant by the trunk); 立 *li,* to stand (a man standing); 降 *chiang,* descend (hill and two footprints pointing downwards); 陟 *chih,* to mount (footprints upwards); 至 *chih,* arrive at (arrow hitting target); 回 *hui,* revolve (meander); 曰 *yüeh,* speak (mouth and breath); 甘 *kan,* sweet (mouth and something in it); 高 *kao,* high (picture of a high building); 長 *chang,* senior, grown up, *chhang,* extended (long-haired man walking on stick); 力 *li,* strength(ard or plough); 富 *fu,* blest (picture of a jar); 酉 *yu,* wine-must (jar and liquid inside).

ASSOCIATIVE COMPOUNDS

父 *fu,* father (hand and stick); 媍 *fu,* wife (女 woman and 帚 broom); 好 *hao,* to love, *hao,* good (woman and 子 child); 奻 *wan,* to quarrel (two women); 林 *lin,* forest (two 木 trees); 森 *sên,* umbrageous (three trees); 析 *hsi,* split (tree and 斤 axe); 牧 *mu,* tend cattle (ox and hand wielding whip); 鳴 *ming,* sing (鳥 bird and 口 mouth); 男 *nan,* male, man (employ 力 strength in the 田 fields).

DETERMINATIVE-PHONETIC CHARACTERS

耳 *êrh,* ear, is PHONETIC in: 珥 *êrh,* ear-pendant (determinative 玉 jade, precious stone; word cognate to 耳); 餌 *êrh,* cake (det. 食 food or 鬲 cauldron); 毦 *êrh,* plume (det. 毛 hair); 佴 *êrh,* assistant (det. 亻人 man); 蛆 *êrh,* bait (det. 虫 worm); 衈 *êrh,* a sacrifice (det. 血 blood); 恥 *chhih,* shame (det. 心 heart); 弭 *mi,* repress, ends of a bow (det. 弓 bow); DETERMINATIVE in: 聞 *wên,* to hear (phonetic 門 *mên*); 聆 *ling,* listen to, apprehend (phon. 令 *ling*); 聾 *lung,* deaf (phon. 龍 *lung*); 聰 *tshung,* acute of hearing, clever (phon. 悤 *tshung*); 愯 *sung,* alarm, excite (phon. 從 *tshung*).

立 *li,* to stand, is PHONETIC in: 笠 *li,* conical hat (det. 竹 bamboo); 粒 *li,* grain of rice (det. 米 rice or 食 food); 苙 *li,* pen for animals, *chi,* hyacinth (det. 艸 herb, plant); 泣 *chhi,* to weep (det. 氵水 water); 拉 *la,* to pull, break (det. 扌手 hand); 翋 *la,* to fly (det. 羽 wings); 霳 *li,chhih,* heavy rain (det. 雨 rain); 颯 飀 *sa,* storm (det. 風 wind); DETERMINATIVE in: 站 *chan,* to stop (phon. 占 *chan*); 竚 *chu,* to wait for (phon. 宁 *chu*); 竣 *chün, tsun,* stop work (phon. 夋 *chün*); 靖 *ching,* quiet (phon. 青 *chhing*); 端 *tuan,* extremity, origin, end, principle (phon. 耑 *chuan*); 竭 *chieh,* exhausted (phon. 曷 *ho*).

Reproduced from G. Haloun, 'Chinese Script', *World Review* (Sept. 1942), by permission; with some modifications.

same. Due to a strong inclination of the Chinese to pun, it led some characters which had ceased to have their original function being used for other purposes. Thus *lai*, meaning to come (來), originally meant a cereal plant, as its ancient ideograph (𤯓) shows, and *wan*, ten thousand (萬), was originally a scorpion (𠂤). The changes happened because the words are homophones, and the new phonetic characters are called *loan characters*.

The greatest invention in the development of Chinese was that of the *determinative-phonetic* characters. A determinative is a basic element (a radical) that is added to a phonetic word to indicate the category in which the meaning of the word is to be sought. Thus a whole series of words with the same, or approximately the same, sound can be written down without any possibility of confusion. Some examples will make this clear. The word *thung*, a phonetic meaning with, together, is combined with various radicals to provide a new series of words:

chin (金) (metal) + *thung* (同) = *thung* (銅) copper, bronze.
chu (竹) (bamboo) + *thung* (同) = *thung* (筒) pipe, flute.
hsing (行) (to go) + *thung* (同) = *thung* (衕) side street.

On the other hand the radical *shui*, water, can be used in combination with another word to show that the word in question has something to do with water. Thus:

shui (水) (water) + *mo* (末) (branches) = *mo* (沫) (froth, foam).
shui (水) (water) + *chha* (叉) (fork) = *chha* (汊) (branching streams).
shui (水) (water) + *mei* (每) (each, every) = *mei* (海) (the sea).

How far combinations like these were the result of ingenuity by scribes in the tenth to seventh centuries B.C. we cannot tell, but many certainly reveal appropriate, even poetical, contexts of thought. Some ideographs can be both phonetic and radical-determinative, like *erh* (ear) (耳) and *li* (to stand up) (竹), as can be seen in Table 2.

Any one of the pictographs or symbols of the classes mentioned could be used as a phonetic, and so render words that sounded the same or, at least, closely similar. But the number of determinatives was not unlimited, since the number of categories required in the primitive stages of a civilisation was not great. As a result the radical-determinative came to be adopted as a convenient way of forming characters; it was already in full use in the ninth century B.C., and codified in 213 B.C. The first great dictionary appeared in A.D. 121, containing 541 radicals. This large number remained in use for some 1200 years, then it was reduced to 360 and finally to 214, the figure in use today.

To anyone with scientific interests approaching Chinese, a helpful analogy is possible if we consider Chinese characters as molecules

Fig. 2. A sound-table from the *Thung Chih Lüeh* of Chêng Chhiao (*c.* A.D. 1150). The words are located on a co-ordinate system, the longitudinal axis of which (reading from right to left) is 'graduated' with initial consonants, while the vertical axis (reading from above downwards) is 'graduated' with vowels and terminal sounds. The longitudinal axis serves also for classification according to musical notes (third row from the top), while the places on the vertical axis are arranged according to the four tones of speech.

composed of permutations and combinations of a set of 214 atoms. There
may be up to seven atoms in one molecule, and the atoms may repeat
– rather as they do in a crystal – with up to three identical ones in a
character as, for instance, in *sên* (undergrowth), where the wood radical
appears three times (森). Admittedly, breaking down phonetics into
their basic radicals is an artificial and somewhat late process. Originally
some phonetics had no connection with the radicals to which they have
become attached by convention and stylisation, so in these cases it is
very difficult to know where in a dictionary certain characters should
be placed. For this reason, lists of characters 'the radical of which is
not obvious' are sometimes given as a separate list. Again, there are some
very complex characters that completely defy dissection simply because
they are ancient pictographs that have become stylised and now belong
to the complex seventeen-stroke end of the list of Chinese characters.
Kuei (龜) (tortoise) is one of these. Nevertheless the atom-molecule
analogy is useful for a great number of characters, and the significance
of an analysis of written signs will become evident later when studying
Chinese scientific terms.

The six classes of characters described were distinguished first by Liu
Hsin and Hsü Shen in the first and second centuries A.D., and have been
under discussion ever since. Called the *liu shu* (the six writings), they
are:

1 *Hsiang hsing* (images, shapes) = pictographs.
2 *Chih shih* (pointing to situations) = indirect symbols.
3 *Hui i* (meeting of ideas) = associative compounds.
4 *Chuan chu* (transferable meaning) = mutually interpretative symbols.
5 *Chia chieh* (borrowing) = loan characters.
6 *Hsing shêng* (picture and sound) = determinative-phonetics.

The last, determinative-phonetics, form the greatest majority of characters. In the great eighteenth-century dictionary *Khang-Hsi Tzu Tien*
only five per cent are pictographs or symbols, the remaining ninety-five
per cent being determinative-phonetics.

Throughout Chinese history there has been a continual pruning and
simplification, and archaic Chinese contained many more sounds than
modern, or even mediaeval, Chinese. Changes in sounds have also
occurred in other Asiatic languages and a study of all these gives helpful
evidence for the way in which news of natural products, ideas and
techniques was disseminated in the past. In Chinese the changes also
had other significance. For instance, by the eleventh century A.D.,
changes had reached such a stage that Ssuma Kuang produced a new
systematic key to sounds in a series of what came to be known as
'rhyme-tables'. These were found of considerable use and were soon
copied (Fig. 2). Incidentally, such tables have a scientific as well as

Table 3. *Sound-combinations in Chinese* FINALS

INITIALS	final alone	Ch-	Chh-	F-	H-	Hs-	J-	K-	Kh-	L-	M-	N-	P-	Ph-	S-	Sh-	Ss-	T-	Th-	Ts-	Tsh-	Tz-	Tzh-	W-	Y-	FINALS
	··	*	·	*	·	·	·	*	·	·	·	·	·	·	·	*	·	·	·	·	·	·	·	·	·	-uo
	·	*	·	*	·	*	*	*	·	*	·	*	·	·	*	*	·	·	·	·	·	·	*	·	·	-ung
	·	*	·	·	*	·	·	·	·	·	·	·	·	·	·	·	·	·	·	·	·	·	·	*	·	-ün
	·	*	·	*	·	*	*	*	·	·	·	*	*	·	*	*	·	·	*	*	·	·	·	·	*	-un
	·	*	·	*	·	*	·	*	·	·	·	*	·	·	*	·	·	*	*	·	·	·	·	·	·	-ui
	·	·	·	·	·	·	*	·	·	·	·	·	·	·	*	·	·	·	·	·	·	·	·	·	·	-uei
	·	*	·	·	*	·	·	·	·	·	·	·	·	·	·	·	·	·	·	·	·	·	·	·	*	-üeh
	·	*	·	*	·	·	·	*	·	·	·	·	·	·	*	·	·	·	·	·	·	·	·	·	·	-uang
	·	*	·	·	·	*	·	·	*	·	·	·	·	·	·	·	·	·	·	·	·	·	·	·	*	-üan
	·	*	·	*	·	·	*	*	*	·	·	·	*	·	*	*	·	·	*	*	·	·	·	·	·	-uan
	·	*	·	*	·	·	·	*	·	·	·	·	·	·	*	·	·	·	·	·	·	·	·	·	·	-uai
	·	*	·	*	·	·	·	·	·	·	·	·	·	·	*	·	·	·	·	·	·	·	·	·	·	-ua
	·	*	·	·	·	·	·	·	·	·	·	·	·	·	·	·	·	·	·	·	·	·	·	·	*	-ü
	·	*	*	·	·	*	·	*	*	*	*	*	*	·	*	*	*	*	*	*	·	*	*	·	·	-u
	*	*	*	·	·	*	·	*	*	*	*	*	*	·	*	*	·	·	·	*	*	·	·	·	·	-ou
	*	*	*	·	·	*	·	*	*	*	*	*	*	·	*	*	·	·	·	*	*	·	*	·	·	-o
	·	*	·	·	*	·	·	·	·	·	·	·	·	·	·	·	·	·	·	·	·	·	·	·	·	-iung
	·	*	·	·	*	·	·	*	*	·	·	·	·	·	·	*	·	·	·	·	*	·	·	·	·	-iu
	·	*	·	·	*	·	·	·	·	·	·	*	·	·	·	·	·	·	·	·	·	·	·	·	·	-io
	·	*	·	·	*	·	·	·	·	*	*	*	*	·	·	·	·	·	·	*	·	·	·	·	*	-ing
	·	*	·	·	*	·	·	·	·	*	*	*	*	·	·	·	·	·	·	·	·	·	·	·	*	-in
	·	*	·	·	*	·	·	·	·	·	·	·	·	·	·	*	·	·	·	·	·	·	·	·	·	-ih
	·	*	·	·	*	·	·	·	·	*	*	*	*	·	·	·	·	·	*	·	·	·	·	·	·	-ien
	·	*	·	·	*	·	·	·	·	*	*	*	*	·	·	·	·	·	*	·	·	·	·	·	·	-ieh
	·	*	·	·	*	·	·	·	·	*	*	·	*	·	·	·	·	·	*	·	·	·	·	·	·	-iao
	·	*	·	·	*	·	·	·	·	*	·	*	·	·	·	·	·	·	·	·	·	·	·	·	·	-iang
	·	*	·	·	·	·	·	·	·	·	·	·	·	·	·	·	·	·	·	·	·	·	·	·	·	-iai
	·	*	·	·	*	·	·	·	·	·	·	·	·	·	·	·	·	·	·	·	·	·	·	·	·	-ia
	*	*	·	·	*	·	·	·	·	·	·	·	·	·	·	·	·	·	·	·	·	·	*	·	·	-i
	*	·	·	·	·	·	·	·	·	·	·	·	·	·	·	·	·	·	·	·	·	·	·	·	·	-erh
	·	*	*	*	·	·	*	*	*	*	*	*	*	*	·	·	·	*	*	·	·	*	·	·	·	-êng
	*	*	*	*	·	·	*	*	*	*	*	*	*	*	·	·	·	*	*	·	·	*	·	·	·	-ên
	·	·	*	*	·	·	·	·	*	*	*	*	*	·	·	·	·	*	*	·	·	*	·	·	·	-ei
	·	·	·	·	·	·	·	·	·	·	·	·	·	·	·	·	·	·	·	·	·	·	·	·	*	-eh
	*	*	·	*	·	·	*	·	*	·	·	·	*	*	·	·	*	*	·	*	*	·	·	·	·	-ê
	*	*	·	*	·	·	*	*	*	*	*	*	*	*	·	·	·	*	*	·	*	*	·	·	·	-ao
	*	*	*	*	·	·	*	*	*	*	*	*	*	*	·	·	·	*	*	·	*	*	·	*	*	-ang
	*	*	*	*	·	·	*	*	*	*	*	*	*	*	·	·	·	*	*	·	*	*	·	·	·	-an
	·	*	*	·	*	·	·	*	*	*	*	*	*	*	·	·	·	*	*	·	·	*	·	*	*	-ai
	*	*	*	*	·	·	*	*	*	*	*	*	*	*	·	·	*	*	*	·	·	*	·	*	*	-a

linguistic interest, since they were laid out on a regular co-ordinate pattern as in a map or a mathematical matrix and so may well have formed a basis for the development of co-ordinate geometry (to be discussed in the next volume).

The pruning of sounds in Chinese has left the language with too few for use, at least as far as developing a Chinese scientific terminology is concerned. Just how few sounds there are may be judged from Table 3, where rather more than half the possible sound-combinations are missing. The four tones of spoken Chinese multiply the sounds available, it is true, but the problem is still serious; for the 49000 characters in the *Khang-Hsi Tzu Tien* have only 412 sounds in the four tones, or a total of 1648 sounds between them. This gives in theory a total of 30 meanings to each sound, and although the situation is ameliorated to some extent due to the obsolete, poetic or highly specialised nature of some characters, it does mean that there has been little leeway for the development of new scientific terms. Western Europeans could draw on Roman, Greek and even Arabic roots, but no such resources were available to the Chinese. More recently, however, new visual combinations of radicals and phonetics have been formed to overcome this difficulty.

The statement made earlier that Chinese may originally have been monosyllabic must be qualified as far as the spoken language is concerned, because two characters with the same meaning are often run together to avoid homophonic misunderstandings. The words *khan-chien* (look–see) and *kan-hsieh* (grateful) are examples of this. In conversation one also finds people making such explanations as *huo-chhê ti huo* meaning 'fire as in fire-carriage' (i.e. a steam engine), a method that occurs frequently too with proper names. Doublet-making offsets to some extent the reduction in the number of sounds, but until the thirteenth century A.D. it was only practised in speech, never in the written word. The historian of science has, of course, to take this into account.

In spite of all its faults, Chinese has one great advantage over European languages for, in spite of stylistic changes, it has remained by and large the same. Anyone who can read the language has relatively little difficulty in understanding a text of any age, whether it was written now or a couple of thousand years ago. This is not so with European languages where the writing has evolved with the spoken word, so that what is virtually a new literary language appears after only a few centuries. And what is more, even if there are some ambiguities, Chinese nevertheless has a concentrated, laconic, lapidary quality that makes an impression of austere elegance, pith and virility, unequalled in any other tongue.

3

The geography of China

The geographical background of China – the stage on which the drama of the development of Chinese civilisation was played – took more than a passive role in determining the differences between the cultures of Europe and China, as will become clear later.

At a first glance, China can be said to be divided laterally by its two main rivers, the Yellow River and the Yangtze (see Fig. 1). This is an oversimplification, but in a country both mountainous and plain with vast deserts and fertile areas, the rivers form a useful network with which to begin a brief physical description. We may conveniently look first to the north-east and the Pei-chih-li gulf. Into its north-eastern shores flows the Liao River which has come down from Manchuria, and opposite, on its south-western shore, is the mouth of the vast Yellow River. Following the Yellow River upstream, on the left is the sacred mountain Thai Shan, once venerated for its rain-giving dragons, and the whole mountainous peninsula of Shantung; on the right is the North China plain. The river now runs through mountainous country, and we turn northwards and then westwards, with the Gobi desert to the north. Next we sail south-westwards to the great city of Lanchow, and to where the river comes down from its source in the Tibetan massif. Rather more than half the area inside this vast curve of the Yellow River is fertile, separated from the Ordos desert approximately by the line of the Great Wall and the romantic city of Yulin, where the sand blows up in drifts as high as the walls and the triumphal gateways.

The Yellow River is not alone – other rivers flow into it: the Chhin, the Lo on the shores of which stood the one time capital Loyang, another river also called the Lo, and the river Wei. The whole south slope of the Wei River valley is a sharp escarpment, but the northern slopes are gentler since the ancient rock formations are covered to a depth of thirty metres or more by yellow loess – a compacted dust blown for long ages southwards from the northern deserts. Along these slopes are the tumulus graves of former emperors, and the whole region is saturated with history; it saw the earliest Chinese civilisation, the rise of the feudal state of Chhin and the successive glories of the Han and Thang capitals at Chhang-an (Sian). This region – the eastern part of Kansu province, the south-eastern part of Ninghsia, and the centre of Shensi – forms a distinct natural province, for in spite of the mountain passes in the west, south and south-east, it is self-contained in a way frequently to be found in Chinese geography.

The Yangtze River, more navigable than the Yellow River, flows into the Pacific north-west of Shanghai. Again, if one travels upstream, one soon reaches a relatively flat country with three lakes, the Thai Hu (Hu = Lake), Poyang Hu, and Tung-thing Hu, the provinces Hupei (north of the lake) and Hunan (south of the lake) deriving their names from the latter. Hopei and Honan similarly mean north and south of the (Yellow) River. At Hankow, between the two western lakes, the Yangtze is joined by the Han River, which flows south, and then the Yangtze itself comes down from Szechuan province through a series of mighty gorges comparable with the Grand Canyon and the African Great Rift valley. On reaching the brick-red sandstone soil of the Szechuan plateau-basin with the Tibetan mountain block to the west, we come upon another self-enclosed natural province which, in World War II, proved an impregnable fortress against the Japanese. Again these central regions were isolated from the south-eastern and southern coasts, which stretch in a huge arc from Hangchow to the Indo-Chinese border. Only comparatively recently have there been good communications in and between them.

The mountains in China seem to be nothing less than a vast jumble of ranges, but in fact they may be broken down into three main groups – north-east/south-west folds, a series of east/west ranges, and a host of lesser folds. The great belt of the north-east/south-west folds is like a series of gigantic steps up to the Tibetan massif (Fig. 3), and they encompass between them most north-eastern and east-central key economic areas. As far as the east/west ranges are concerned, the four main ones effectively divide up the country into four domains – the Shensi basin (west of Loyang), the red plateau-basin of Szechuan, the Kweichow plateau (west of Honan), and the southern maritime area, especially around Canton. The other folds contain the so-called 'Yunnan arc' which the upper Yangtze follows as it comes down from the Tibetan massif.

Essentially, then, China – and Central Asia too, for that matter – are distinguished from most other regions by having a complex network of high mountain ranges separating a number of flatter areas. The

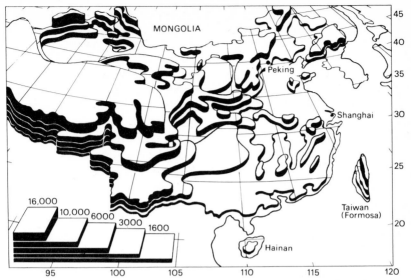

Fig. 3. The eastward staircase of the Chinese subcontinent. Based upon Lo Kai-Fu, 'The Basic Geography of China', *China Reconstructs*, 1956, **5** (no. 12), 18.

significance of this is that, compared with Europe, China is far more cut up by mountain ranges into isolated regions not accessible by inland seas, or even by simple roads. It was, then, an immense task to try to unify Chinese culture and stabilise a common language.

Besides a geography of wide contrasts, China is also subject to a great range of climatic conditions. In the west, the Tibetan massif is mostly an inhospitable waste of high frozen desert with mainly nomadic inhabitants tending yak and goats. The Tarim basin and Mongolian steppe to the north are a mixture of arid grassland and true desert, beautiful but sparsely inhabited. Life here is also traditionally nomadic and never coalesced with the mainstream of Chinese civilisation until our own time. The grassy steppes of the Manchurian plain, in spite of being frost-free for only five months of the year, are now well developed agriculturally, although of minor importance historically.

Moving down from the arid Shantung mountains in the east, however, we at last enter historic ground – part of the ancient feudal state of Chhi – where, in spite of severe winters, it is possible to grow some staple crops and produce a kind of wild silk based on the oak, not the mulberry. The adjoining North China plain to the south-west is a great area of alluvial soil, fertile given intense manuring (including human excreta). Composed of silt carried down by the Yellow River, it teems with life in spite of periodical floodings, and many crops are cultivated; the chief is wheat but there are also millet, beans, cotton and hemp.

To the north and north-east, the Shansi, Shensi and Kansu provinces show a totally different picture. Covered with loess giving an unleached soil of exceptional richness, crops can be raised year after year without fertilisation and the ability of the soil to hold moisture allows good harvests in spite of limited rainfall. Here, in an area particularly favourable to fruit-growing, was the oldest focus of ancient Chinese agriculture. Different again is the Lower Yangtze valley to the east. With a network of canals and rivers, no land is unused in what is the heart of rice-growing country. Rice, too, is cultivated south of the southern borders of the Shantung, Honan and Shensi provinces, that is in the south-eastern maritime provinces. But here the main activities are lumber and fishing; Chinese sailors are traditionally Fukinese and Cantonese. The climate is warm and wet, and the language composed of mutually incomprehensible dialects, a result of the area's historic isolation from the interior.

In the west lies the red basin of Szechuan, one of China's most densely populated, attractive, and fruitful regions. Rice is the main crop, but there are many others, including cotton, sugar-cane, oranges and tobacco, and winter crops as well, some farmers harvesting no less than three crops a year. South of the area are the Kweichow and Yunnan highland plateaux and mountain blocks. Here only some five to ten per cent of the region has flat ground, but the climate has the moderate and agreeable character typical of places at high altitudes in tropical and subtropical latitudes, and where flat ground does exist, it is intensely cultivated.

Lastly, to complete this brief impression of China's wide-ranging geographical conditions, there are the seaward-looking valleys of Kuangtung and the plateau of Kuangsi. Both Kuangs are subtropical with long hot summers of high humidity, and rather cool winters that are followed by two transitional months of mist and fog. Farming is predominantly centred on rice cultivation, but there are also crops of sugar-cane, tobacco and oranges, as in Szechuan. Three crops a year are common, and much silk is also made in the region.

4

Chinese history:
(i) The pre-Imperial Ages

China is better provided with original source-material about its past than any other Eastern, and indeed most Western, countries. Unlike, for example, India, where chronology is still very uncertain, with China it is often possible to be certain not only of the year, but even of the month and day as well. A great number of official histories and annals have survived, all written with remarkable lack of bias, but it is an unfortunate fact that only very small parts of them have been translated into European languages. They are very valuable from an economic, political and social point of view, but generally speaking of relatively little use for the history of science. They do indeed provide much astronomical and meteorological information because the heavens were the fundamental basis for computing calendars, and events in them, as well as the weather, were used to foretell the future. But Chinese literary culture was on the whole uninterested in science, and by and large the historian of science has to look elsewhere for his evidence. Happily a vast amount of information is available: it is to be found in what the Confucian scholars classified as 'Miscellaneous' writings, and though few have been translated, Joseph Needham and his collaborators have been in a position to draw on them extensively.

We shall, of course, need a framework of historical dates against which scientific discoveries may be set, and to do this means compiling a list of dynasties, since the Chinese, like the mediaeval European, counted time from the accessions of kings or emperors. Chinese dynasties, however, do not follow a completely consecutive course, but Table 4 should orientate us into their long time-scale of close on four thousand years. Admittedly, there is some uncertainty when it comes to dates prior to 841 B.C., but after that there is general agreement and, fortunately for us, the history of science is concerned primarily with the later period.

Table 4. *Chinese dynasties*

夏 HSIA kingdom (legendary?)		c. −2000 to c. −1520
商 SHANG (YIN) kingdom		c. −1520 to c. −1030
	Early Chou period	c. −1030 to −722
周 CHOU dynasty (Feudal Age)	Chhun Chhiu period 春秋	−722 to −480
	Warring States (Chan Kuo) period 戰國	−480 to −221
First Unification 秦 CHHIN dynasty		−221 to −207
	Chhien Han (Earlier or Western)	−202 to +9
漢 HAN dynasty	Hsin interregnum	+9 to +23
	Hou Han (Later or Eastern)	+25 to +220
三國 SAN KUO (Three Kingdoms period)		+221 to +265
First Partition	蜀 SHU (HAN)	+221 to +264
	魏 WEI	+220 to +265
	吳 WU	+222 to +280
Second Unification 晉 CHIN dynasty: Western		+265 to +317
Eastern		+317 to +420
劉宋 (Liu) SUNG dynasty		+420 to +479
Second Partition Northern and Southern Dynasties (Nan Pei chhao)		
	齊 CHHI dynasty	+479 to +502
	梁 LIANG dynasty	+502 to +557
	陳 CHHEN dynasty	+557 to +589
	Northern (Thopa) WEI dynasty	+386 to +535
魏	Western (Thopa) WEI dynasty	+535 to +556
	Eastern (Thopa) WEI dynasty	+534 to +550
北齊 Northern CHHI dynasty		+550 to +577
北周 Northern CHOU (Hsienpi) dynasty		+557 to +581
Third Unification 隋 SUI dynasty		+581 to +618
唐 THANG dynasty		+618 to +906
Third Partition 五代 WU TAI (Five Dynasty period) (Later Liang, Later Thang (Turkic), Later Chin (Turkic), Later Han (Turkic) and Later Chou)		+907 to +960
遼 LIAO (Chhitan Tartar) dynasty		+907 to +1125
West LIAO dynasty (Qarā-Khiṭāi)		+1124 to +1211
西夏 Hsi Hsia (Tangut Tibetan) state		+986 to +1227
Fourth Unification 宋 Northern SUNG dynasty		+960 to +1126
宋 Southern SUNG dynasty		+1127 to +1279
金 CHIN (Jurchen Tartar) dynasty		+1115 to +1234
元 YUAN (Mongol) dynasty		+1260 to +1368
明 MING dynasty		+1368 to +1644
清 CHHING (Manchu) dynasty		+1644 to +1911
民國 Republic		+1912 to +1949
People's Republic		+1949

NOTE. When no modifying term in brackets is given, the dynasty was purely Chinese. During the Eastern Chin period there were no less than eighteen independent States (Hunnish, Tibetan, Hsienpi, Turkic, etc.) in the north. The term 'Liu chhao' (Six Dynasties) is often used by historians of literature. It refers to the south and covers the period from the beginning of the third to the end of the sixth centuries A.D., including (San Kuo) Wu, Chin, (Liu) Sung, Chhi, Liang and Chhen. The minus sign (−) indicates B.C., and the plus sign (+) is used for A.D.

Chinese pre-history and the Shang dynasty

The first inhabitants on Chinese soil, of whose remains we know, were the race to which 'Peking Man' belonged. *Sinanthropus pekinensis* lived at the beginning or middle of the Pleistocene period, that is around 400000 B.C., and earlier than Neanderthal Man of Europe and the Mediterranean. There is certain evidence, too, that there was a Neolithic population in China around 12000 B.C., but after this there is a remarkable gap in continuity; only in Manchuria are all the subsequent prehistoric stages to be found. Then, suddenly, about 2500 B.C., the apparently empty land begins to support a large and busy population (Table 5). There are hundreds, even thousands, of villages, inhabited by a people tending flocks and having an agricultural economy, acquainted with textiles, carpentry and ceramics. Obviously, extensive archaeological work is needed to throw light on this curious hiatus between the Stone Age and their late Neolithic successors.

The first important culture in China to be revealed by excavations is the Yangshao, which existed in a belt of country running from west to east, and comprising the present provinces of Kansu, Shensi, Shansi, Honan and Shantung. The chief cereal was almost certainly millet and, later, rice, and since neither plant is indigenous to China, the likelihood is that they came from South-East Asia. Bones of dogs, pigs and, later, sheep and cattle, have been found; bones of the horse have been identified too, although this may have been a wild horse of the kind that until recently still existed in Mongolia. The most outstanding characteristic of the Yangshao culture, however, was their painted pottery, made about 2500 B.C. by coiling clay, not by using a potter's wheel.

It is worth noting that at this stage of development, there is evidence for a wide community of culture throughout Northern Asia and Northern America. For instance, an implement to be found throughout the area was a rectangular or semi-lunar knife, quite unlike anything in Europe or the Middle East, but used by Eskimos, Amerindians, Chinese and in Siberia, and this as well as other evidence points to a migration across the Bering Straits. On the other hand, the Chinese themselves developed some characteristic inventions of their own at this time, most notably two types of pottery vessel which seem to be unique, and are of interest to the historian of science in view of the close connection (to be explored in a later volume) between cooking and chemistry. One was the *li*, a cauldron or pot with three hollow legs which increased the surface area to be heated and so led to greater efficiency (Fig. 4). The other was a *tsêng*, a vessel with a perforated bottom that could be stood

Table 5. *Some historical chronology*

Dates	Mediterranean	Egypt	Palestine	Mesopotamia	India	China
B.C. 3500				Beginning of the city of Ur		
	Early Minoan (c. 2600 B.C.)	Old Kingdom (2600 B.C.) Pyramid building (2700–2500 B.C.) Middle Kingdom (2100 B.C.)			Indus valley civilisation	Yangshao civilisation
2500				Gudea of Lagash (2100 B.C.)		
2000	Middle Minoan (c. 1800 B.C.)		Abraham (?)			Lung Shan civilisation (1600 B.C.) Shang dynasty
1500		New Kingdom Tutankhamen (1361–1352 B.C.)	Moses (c. 1300 B.C.)			Chou conquer Shang (1027 B.C.)
1000			David (c. 1000 B.C.) Solomon (c. 950 B.C) Destruction of temple at Jerusalem	Zoroaster (in Iran) (c. 600 B.C.) Nineveh destroyed (c. 600 B.C.) Fall of Babylon to Cyrus (538 B.C.)	Guatama (Buddhism) (c. 560 B.C.) Mahāvīra (Jainism) (c. 560 B.C.) Invasion of Punjab by Darius (512 B.C.)	Confucius (c. 550 B.C.)

Date			
500	Warring States (480 B.C.)		
400	Erection of Parthenon at Athens (450 B.C.)		
	Plato (428–348 B.C.)		
	Aristotle (384–322 B.C.)		
	Conquests of Alexander the Great (c. 327 B.C.)		
300		Reign of Asoka (300–274 B.C.)	
200	Punic Wars in the Mediterranean (250–150 B.C.)		Supremacy of the Chhin (212 B.C.)
			Han dynasty (202 B.C.)
100			
	Cleopatra (69–30 B.C.)	Roman capture of Jerusalem (63 B.C.)	
A.D. 0	Beginning of Roman Empire (31 B.C.)		
100		Destruction of Jerusalem (A.D. 70)	
200	End of Han (A.D. 220)		

Fig. 4. A Shang dynasty *li* (see p. 21). This example dates from the seventeenth to sixteenth centuries B.C.; its height is 16.5 cm. Photo © Times Newspapers Ltd., by permission of Robert Harding Associates.

Fig. 5. A *hsien* of the Shang dynasty. This is a pottery version of the combination steam cookery vessel (see p. 26), and was excavated in 1953 at Chêng-chou in Honan. Standing 40 cm high, it is of similar age to the *li* shown in Fig. 4. By permission of Robert Harding Associates.

on top of a *li*, thus making an efficient steam cooking combination, which also had the advantage of allowing more than one foodstuff to be cooked at the same time. When the two were combined into a single form, often with a removable grating, it was called a *hsien* and became more preponderant when bronze replaced pottery (Fig. 5). We now know that this led directly later on to the characteristic East Asian apparatus for distillation.

The Yangshao was followed in Honan and Shansi, by a later Neolithic culture, the Chhêng-Tzu-Yai or Lung-Shan (archaeological site names). Although still without metal, the people of this culture used a smooth black earthenware of fine texture and high finish, while those of Lung-Shan had domesticated all the animals of the Yangshao including, probably, the horse. There is a possibility, too, that the Lung-Shan knew of wheeled vehicles, although the evidence for this is uncertain. This was also the time at which various inventions appeared, inventions such as the potter's wheel and the use of tamped earth for buildings, long known in the Middle East but new to China.

With the Lung-Shan we reach 1600 B.C. and then, within a century, we come quite suddenly to a mature bronze-age culture, the Shang dynasty (Fig. 6). Most of our knowledge comes from excavations at the capital, Anyang, now in Honan province, and the discovery of the very existence of this dynasty is one of the most romantic in all archaeology. It all began in the late nineteenth century when farmers tilling their fields near Anyang kept turning up curious pieces of bone. Bought up by someone in the village, they were sold to drugstores as 'dragon-bones' for medicine, but not for long as in 1899 some Chinese scholars came across them and realised, with surprise, that the bones were inscribed with very ancient writing; by 1902 their full significance was appreciated. They were nothing less than inscribed oracle-bones, and provided evidence that no one had possessed since the early days of the Han dynasty (Fig. 7). At one stroke they pushed back the philology, linguistics and history of China by almost a thousand years, and have made it clear that much of China's hitherto legendary history – the rule of the Yellow Emperor, the work of the Great Engineer Yü, and a host of others – was a reflection of events and practices of times that had been in fact historic. One aspect of this is the question of what script preceded the well-developed writing on the oracle-bones. Recently many signs have been recorded from Neolithic pottery, and these may have been the earliest forms of what afterwards developed into the Chinese characters.

The oracle-bones were used for scapulimancy, a divination technique that involved heating the shoulder-blades of mammals or the shells of turtles with a red-hot poker and discerning the reply of the gods from

Fig. 6. A magnificant Shang dynasty ritual wine vessel in bronze. Excavated in 1957 at Fu-nan, Anhui, it dates from the fourteenth to eleventh centuries B.C., and stands 47 cm high. By permission of Robert Harding Associates.

the shapes and directions of the cracks. It seems to have been a method peculiar to the Anyang area and to have started just a little before the arrival in 1520 B.C. of the Shang dynasty, when the technique was developed. Indeed, the Shang diviners were so well organised that they

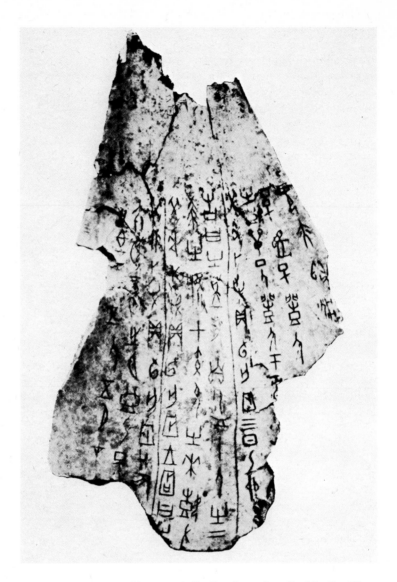

Fig. 7. Oracle-Bone (Shang period). Specimen described by Lo Chen-Yü. Reproduced from L. C. Hopkins, 'Sunlight and Moonshine', *Journal of the Royal Asiatic Society*, April 1942, by permission.

kept records of their results, possibly as secret dossiers; it is a collection of such records that the Anyang finds have brought to light.

The use of bronze during the Shang dynasty was outstanding. It was employed in all kinds of ways, especially for ritual, military and luxury purposes, for the metal parts of wheeled vehicles, but, interestingly enough, seldom for tools and implements. The high artistic quality and workmanship of Shang bronze ritual vessels is breathtaking and surpasses all later work. Yet the Shang ruled over only a restricted area, perhaps spreading no more than 300 km in any direction from Anyang. Theirs was a feudal society in which matriarchal traces had given place to patriarchal control, where there was family or ancestor worship, and human sacrifice. Slaves were immolated at royal burials, a practice that persisted well into Chou times.

Two other features of the Shang Age are worth noting here. One is the extensive use of bamboo, not least made up into books for writing. Probably similar to the Han books that still survive, bamboo strips were taken and held together by two lines of cords, and it is from these that the character *tshê* (╪╫), depicting a written book, was derived. Incidentally, it was during Shang times that the Chinese writing brush was introduced, and that pictographs began to be replaced by written characters. The second feature is the use of cowrie shells as a form of currency, an innovation that gave many words expressing 'value' the radical *pei*, which originally meant cowrie shell. Where the cowries came from is still uncertain: the Pacific coast south of the Yangtze estuary seems probable, but their journey to the centre of the Shang civilisation would present a remarkable feat.

The Chou period, the Warring States and the first unification

We turn now to the Chou people who came from the western regions (the present provinces of Kansu and Shensi). Less advanced than the Shang, whom they admired, the Chou conquered the Shang area about 1027 B.C., encouraged the bronze-working, pottery and textile-making they found there, and developed the written language.

Though probably of pastoral antecedents, the Chou soon adopted the thorough-going agricultural character of the unfolding Chinese civilisation. In addition they systematised the incipient feudal system of the Shang until it became almost as developed as it was to become two millennia later in Europe, basing their economy on the work of peasant-farmers who all had to do their quota of unpaid labour on the land of the local nobility. The empire, as it had now become, was divided in fiefs, held by a new aristocratic class, and much of the Shang population was deported to the dukedoms of Lu and Chhi.

In spite of its apparently overriding control, Chou society began to manifest an increasing instability which, in the eighth century B.C., led to the collapse of the myth that it was a tightly knit feudal empire. In 771 B.C. the Emperor Lu was killed by an army of one of the smaller states which had allied itself with 'barbarians', and his successor was forced to move the state capital from near Sian eastwards to Loyang and vacate the fertile loess region to the west. During the next few centuries some twenty-five feudal states, paying only lip-service to Loyang, vied with one another in attempts to gain independence. The first to achieve it was the state of Chhi in Shantung, a state which had two peculiar qualities of its own: it was the main source of salt (obtained by evaporation of sea-water) and a leader in the working of iron. Iron had become known in China by 500 B.C. and the Chhi possessed an enviable iron-working technology that the Chou did not, a factor that may well have been significant in their bid for power. Later, independence also passed to the states of Sung, Chin, Chhin and Chhu.

By the sixth century B.C. not only do we see far-reaching political changes, but we also enter the greatest period of intellectual development in ancient China. The 'hundred schools' of philosophers were at their height between 500 and 250 B.C., with scholars travelling with their disciples from capital to capital to act as advisers to the feudal lords, whose realms were beset by conflicts with barbarians, unrest within and vast technological changes due to the increasing use of iron. During this period academies of scholars were set up, the most famous being the Academy of the Gate of Chi (Chi-Hsia) at the Chhi capital, founded in 318 B.C. by Prince Hsüan. Here scholars were welcomed from other states besides Chhi, and all were provided with quarters and maintenance. Established not long after Plato's Academy in far-off Athens, it attracted a great number of superb scholars, some of whom we shall meet later.

Side by side with these intellectual developments went a host of other advances, so this has always been looked on as China's 'classical' period. There were developments in craftsmanship, in methods of production, and in irrigation: the animal-drawn plough appeared, there was a multiplication of market-places and an intensification of a money economy which tended to replace the ownership of land and of unpaid labour duty as a source of wealth. In military techniques iron came to be widely used, and the cross-bow, invented earlier in China than elsewhere, brought to a state bordering on perfection.

During the time of the Warring States, industrial concentration and the control of hydraulic engineering systems played a great part in what was a general consolidation, with the smaller states being absorbed by

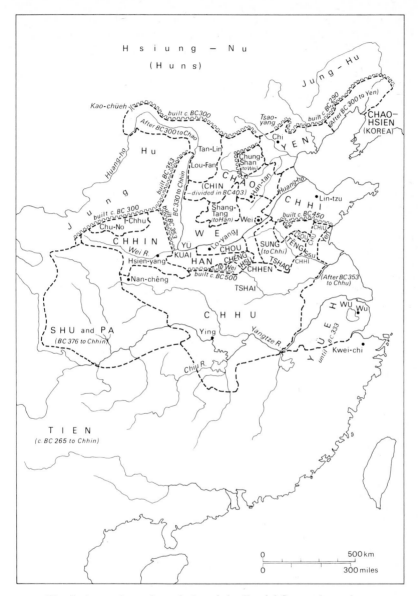

Fig. 8. Approximate boundaries of the Feudal States about the beginning of the third century B.C. Based upon A. Herrmann, *Atlas of China*, Harvard University Press, Cambridge, Mass., 1935, by permission.

the larger. Feudalism was replaced by bureaucratism, first in the state of Chhin and then elsewhere. The population became increasingly militarised, a police and passport system was introduced, and general coercion exercised by the use of drastically severe punishments. The power of Chhin continued to grow and was viewed with alarm by other states, which made alliances in self-defence and adopted other expedients to try to increase their security. Yet try as they would, they were unsuccessful, as for instance when a number banded together and approached the Chhin with proposals for a vast inland navigation system. The idea was to build a canal connecting the Ching and Lo rivers that lay to the north of Sian, in the belief that constructing the canal would keep the population occupied and reduce Chhin military strength. The result was just the opposite; a vast area became irrigated, additional grain was grown, with the Chhin realising that they now had the wherewithal to build up a greater army and become strategically stronger. Indeed the Chhin found the scheme so satisfactory that the construction of large water-works became a settled policy of their administration, and in 316 B.C. they undertook the magnificent project of irrigating the Chhêngtu plain, a project that still flourishes today.

For almost a century, from 318 to 222 B.C., the Chhin adopted a policy of conquest until, at last, they ruled a united China, and Prince Chêng adopted the title Chhin Shih Huang Ti, the First Emperor of a united China (Fig. 8). Yet this triumph of the Chhin was short-lived, and to its demise and the rise of its successor, the Han dynasty, we now turn.

5

Chinese history: (ii) The Empire of All Under Heaven

The dynasty of Chhin

As soon as their new unified empire was established, the Chhin created a type of bureaucratic government that was to form the pattern for all subsequent Chinese history. The large feudal estates were expropriated and managed by government officials, such nobles as remained being forced to reside at the capital; farmers were granted more rights over their lands than ever before, but they were made subject to taxes. Administratively the country was divided up into provinces, each with a military governor and a body of civil officials, and there was wide-scale standardisation, ranging from weights and measures to the gauge of carts and chariots. Tree-lined roads, already begun in some provinces, were extended into a network, and, in the north, joined together to form a supply route to the immense defence structure of the Great Wall, the dividing line between the steppe and the cultivated land. A military device to prevent incursions from the nomadic tribes, it could only be breached either through one of its heavily defended gates or by building ramps to permit the wall itself to be scaled; in either case there would be a long enough delay for reinforcements to arrive. Yet the purpose of the Great Wall was not only to keep out barbarians from the north; it was also designed to act as a barrier to migration from China itself, to stop any movement out to the north and the formation of mixed agricultural and pastoral economies (Fig. 9).

The Chhin retained a large standing army which had, of course, to be kept busy; an aim which the Emperor Chhin Shih Huang Ti achieved by a series of campaigns to extend the borders of his empire further south. These campaigns were remarkable in their extent, covering the coastal province of Fukien, the two Kuang provinces, and even penetrating as far as Tongking in what is now North Vietnam. One of these, consisting of 3000 likely young men, artisans, and useful girls, was headed by Hsü Fu, whose tomb can still today be seen at Shingū

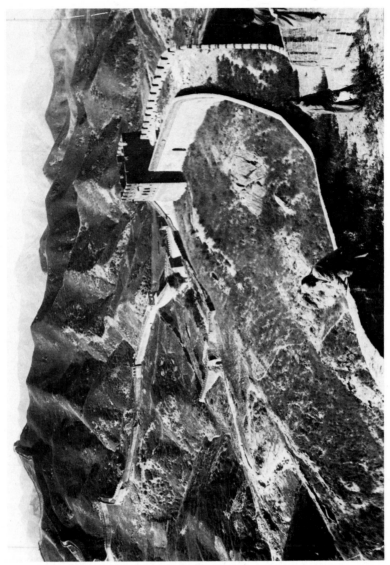

Fig. 9. The Great Wall near the Nankow Pass, north of Peking (photograph of about 1910: Ponting).

in southern Honshu. That they actually got to Japan and settled there, contributing largely to the population and its arts, appears from very recent evidence in dental archaeology. The teeth of Japanese skulls are similar in many ways to those of Shang Chinese skeletons, while those of the Jomon people, earlier inhabitants of Japan, resemble those of the modern Ainu, now living only in Hokkaido and other northern parts of the country. Yet the emperor was not solely interested in military exploits or purely governmental matters; under the guidance of his minister Li Ssu, he standardised the Chinese language, and took a lively interest in alchemy as well as magic. He was also, it is said, an indefatigable worker, dealing with over 1.5 tonnes of reports on wood and bamboo tablets a month and travelling widely over the whole of his empire.

But when in 220 B.C. the Great Unifier died, his son proved an ineffectual ruler and the empire began to crumble. Trouble came first from a 'Back to Feudalism' movement, with some of the former feudal states setting up their own administrations once again. Then there followed a struggle for the throne within the Chhin, a struggle won by the army commander Liu Pang whose rise to power was, in a sense, a study in desperation. Originally in charge of a group of convicts, Liu Pang was subject to a death sentence after some of the group managed to escape; with nothing to lose he therefore deserted and became head of a gang of bandits. When finally the opportunity came, Liu Pang seized it, taking control of the entire government in 202 B.C., and adopting the name Han Kao Tsu. Liu Pang thus founded the Han dynasty which proved to be stable, and was to last for the next 400 years.

The dynasties of Han

However much some might want to return to feudalism in its original form, this was seen to be completely impractical, and Liu Pang settled for a compromise. A number of small feudal states were confirmed but only within a wider state framework based on Chhin lines, and with stringent laws on succession and government. Whenever the ruler of such a small state died, the state was divided, and should a ruler take any false steps in government, then this was also made the excuse to reduce the state's territorial boundaries. In addition, an official from the central government resided at each vassal court. The end was inevitable: the power of every petty state dwindled away until China was once more under the control of a powerful central government.

Liu Pang established his capital at Chhang-an (modern Sian) and divided the country into thirteen provinces. A large bureaucracy was needed and the manpower recruited by competitive examinations, a method that gave the Confucians (to be discussed later, Chapter 7) a hold on Chinese society that was to become permanent. The Confucians were able to achieve this power partly because of the unpopularity of the previous advisers, the Legalists, whose measures were now considered to have been too rigorous, and they confirmed themselves on a sound footing by introducing simpler laws and less severe punishments. Indeed, during Han times the Confucian bureaucracy turned into a highly developed form of civil government. Assemblies of scholars were held to determine a kind of case law and decide on precedents based on ancient writings, the first such assembly, held in 51 B.C. in the Shih-Chhü (Stone Canal) Pavilion of the Palace, having an importance in Chinese history on a par with the Council of Nicaea (A.D. 325) in Western Christendom.

It was during Han times that eunuchs began to exert an increasing power at court. Although condemned by the scholar-gentry of the civil service, they always managed to reassert their influence after every setback, due partly to the fact that no ruler need fear a eunuch establishing a dynasty of his own. This was a valuable asset in a bureaucracy where it was customary to try to make offices hereditary and where nepotism often led to bloody revolutions every time there was a change of reign. The power of the eunuchs reached one of its peaks at the time of the great Han Emperor Wu Ti because, unlike his predecessors, he was unwilling to leave his ministers to govern, and took things into his own hands. As a result the bureaucrats lost their position as intermediaries between emperor and people, and in the end it was only the eunuchs who exercised any power, since it was they and they alone who had access to the private apartments and could there gain the emperor's ear.

Han Wu Ti's reign (from 140 to 87 B.C.) was one of the most important epochs of Chinese history, for it brought stability and a well-administered state with an enlightened foreign policy. Yet there were economic difficulties to begin with; previous anti-mercantile edicts had driven the merchants to wild speculation and sent prices so high that additional currency had been required. Wu's counter-measures were simple and effective: he persuaded the most capable merchants to enter the administration, and conducted currency experiments, trying out the first 'paper' money, which, made from the skin of white deer to be found only in the royal hunting reserves, was used in specific cases of compulsory purchase by the state. As an additional stabilising force, the conception of the 'ever normal granary' was introduced: here the government purchased grain when prices were low and sold it when prices rose. Nevertheless continuing struggles against the Huns north of the Great Wall kept what were still high taxes on the increase.

Han contacts with other countries led to several maritime visits from Romans and Roman Syrians, but is perhaps best exemplified by the extraordinary diplomatic mission of Chang Chhien. A yeoman, he was sent westwards some 5000 km to the Yüeh-chih in Bactria (now an area of North Afghanistan and parts of Tadzhikistan and Uzbekistan in the south of the U.S.S.R.). The purpose of his mission was to get the Yüeh-chih to ally themselves against the Huns, the emperor believing there was a good chance of success because the Huns had killed their king and further insulted the Yüeh-chih by using his skull as a drinking vessel. Unfortunately Chang Chhien was captured by the Huns both on his outward journey and his return, and detained by them for some ten years. Yet the fact that he was able to undertake the mission at all and return safely with a valuable collection of plants and natural products, after twice using a route that took him through enemy territory, is nothing less than remarkable. However, it had more importance than an epic adventure, for Chang Chhien's visit to Bactria led to the later expansion westwards of the Chinese Empire, with the foundation of the famous trade-route that linked the culture-areas of China and Iran, known as the Old Silk Road. Above all, it was the discovery of Europe by China, not the reverse, for Bactria had been Greek since Alexander's time.

The Han emperors, and Wu Ti in particular, are sometimes criticised for 'superstitious practices'. Certainly Wu extended sacrifices and magico-religious rites and spent considerable time in trying to cultivate relationships with spiritual beings, but he was too acute to be fooled by false manifestations even if he could not bring himself to believe that everything his magicians did was fraudulent. It may well be that he was not far wrong in his assessment of them for we know that in earlier ages there were intimate connections between magic and science, and it is highly probable that the Han magicians did make some worthwhile new observations in alchemy, magnetism, the use of medicinal herbs, and so on. Indeed this, the second century B.C., was the time when the greatest classics of Chinese medicine were compiled, especially the *Huang Ti Nei Ching* (The Yellow Emperor's Compendium of Corporeal Healing). This corresponds in large measure to the slightly earlier Hippocratic Corpus of the Greeks, and has been the foundation of Chinese medical thought during the subsequent 2000 years. Among other things it contains the first systematisation of the acupuncture technique.

Emperor Wu, like Huang Ti before him, sent maritime expeditions into the Eastern Ocean, in the belief that spiritual beings dwelt on some of the Pacific Islands. He and his successors also mounted more warlike excursions to the south, and to Korea in the north-east where a colonial government was set up, a government which, in its turn, exerted a great influence on the slowly developing culture of Japan.

From A.D. 9 to 23 the Han dynasty suffered an interregnum, the Regent Wang Mang outmanoeuvring the members of the Han family and establishing himself as the first (and last) Hsin emperor. Wang's rule may have been brief, but it was marked by a series of fundamental reforms that seem to have been aimed at strengthening the bureaucratic state. All land was declared state property, large holdings were distributed among peasant-farmers, and a tax imposed on all uncultivated fields. A law freeing male slaves was also promulgated, but it proved impossible to enforce, and a heavy tax was placed on all slave-owners instead. All gold coins were called in and replaced by bronze, a move that brought immense wealth to the Treasury, which gradually accumulated more gold than was ever to be available in mediaeval Europe. The ever-normal granary system was tried again but like the rest of the reforms it did not work well. Perhaps an honest civil service might have made a success of the new system, but the Wang administration suffered from corrupt officials, drove merchants and financiers to desperation, and made the population restive. Finally, aided by the 'Red Eyebrows' – a secret society typical of those that were so often to play a great role in Chinese society – there was a popular revolt; Wang Mang's power collapsed and he was assassinated.

Wang Mang's reign was not as much a complete fiasco as this brief sketch might seem to imply, because he encouraged the development of the science and technology of his day and, as we shall see in a later volume, may have been intimately concerned with the origin of the magnetic compass. It was Wang, too, who in A.D. 4 convened the first assembly of scientific experts in Chinese history: though unfortunately no record of their deliberations has reached us. Again, fifteen years later, when one man in thirty of the population had been conscripted to fight the Huns, he enlisted experts who claimed to be able to provide scientific and technical assistance to the army. The fact that Wang's tests proved that none of their schemes were practicable is beside the point: it was his attempt to apply such methods that matters.

A short period of confusion followed Wang Mang's assassination, then Liu Hsiu, a cousin of the former Han emperors, emerged triumphant. Forming the Later or Eastern Han dynasty in A.D. 25, he moved the capital eastwards to Loyang and consolidated Han practices and policies. Wars with the Huns continued, but by 80 Pan Chhao, the Han governor-general in Central Asia, had quelled the whole Tarim basin (modern Sinkiang province), and made Chinese influence felt as far west as the Caspian Sea. Only Parthia (modern north Iran), through which the silk route ran, separated the Chinese and Roman empires. After 120, further mercantile contacts were made, mainly with Arabia and Syria by way of the Persian Gulf.

The time of the Han dynasty, especially the Later Han, was one of the relatively important scientific periods in Chinese history. There were great advances in astronomy, improvements in the calendar, an outstanding development in the earth sciences, and foundations laid for methods of classifying plants and animals; alchemy flourished, and the first book ever written on the subject appeared (A.D. 142). A sceptical and rationalist way of thinking developed, particularly about A.D. 80 in the hands of Wang Chhung (see p. 203), while there were two Han princes who also took part in this active intellectual life. One, Tê of Ho-Chien, was a scholar and bibliophile who preserved the important 'Artificer's Record' section of the *Chou Li* (Records of the Rites of Chou), the other was the almost legendary Liu An of Huai-Nan, who gave his name to the *Huai Nan Tzu*, a compendium on all the science of the day and one of the most important monuments of ancient Chinese scientific thought. Indeed, bibliography as a whole received great stimulus, for the Han period marked the first systematic development of book lists; compiled by experts in astronomy, medicine, military science, history, magic and divination, these were incorporated in the Han histories and list some 700 works written on wooden or bamboo tablets, and on silk. Buddhism also entered China in Later Han times and the first sutras were translated into Chinese at the capital, Loyang.

In technology the Han age was marked by the invention and spread of the use of paper, by numerous developments in ceramics such as the first glazes and the introduction of a material that was the forerunner of porcelain, by advances in architectural techniques such as making decorated bricks and tiles, and by raising the level of textile technology to a stage not approached by Iran or Europe until centuries later. A large number of natural products new to China were also imported: alfalfa and the grape-vine from the west, oranges, lemons, betel nuts and lychees from the south and south-west. From the west also came improved breeds of horses, and from Khotan, possibly from Burma too, jade arrived in large quantities. Perhaps the greatest achievement of the Han people in nautical technology was the cardinal invention of the axial rudder at least as early as the first century A.D.

Towards the end of Later Han times, palace revolutions became increasingly frequent, and in 184 a farming crisis led to a peasant revolt guided, in this case, by the 'Yellow Turban' secret society. Although the revolt was suppressed, it left some of the army generals in positions

of great power, and by 220 the central government found itself in-
effective. The country became divided, and for the next half century
remained fragmented into three independent kingdoms in a state of
permanent mutual hostility.

The San Kuo (Three Kingdoms)

The three kingdoms were the Wei, the Wu and the Shu. The Wei
controlled the north and north-west, being based essentially on the
Yellow River valley with their capital at Loyang; the Wu in the south
and south-east ruled the Yangtze valley and the two Kuang provinces,
while the Shu were based on the Szechuan basin in the east, but also
commanded the hills of Kweichow and part of Yunnan (Fig. 10).

The battles and manoeuvres of the Three Kingdoms have become
legendary, inspiring one of China's most famous novels as well as many
plays, with the Wei leader Tshao Tshao as the model of a brave but
cunning and ruthless prince. More important from our point of view,
however, is the fact that the Three Kingdom division was essentially
an economic one, each kingdom covering a key economic area. This
was significant, for in a civilisation based on intensive agriculture, the
accumulation of grain at the centre of power was vital, and in third-century
China this depended on efficient hydraulic engineering, both for irrigation and transport: political power was therefore closely connected
with technology and efficient administration. Regional geography also
played a part, but as the three economic regions were essentially equal
in their natural resources, hydraulic engineering became the main factor
in the power struggle. The Wu completed an important canal and made
an artificial lake for irrigation at Tan-yang (close to modern Nanking),
the Shu carried out some works in the upper Wei valley, but it was the
Wei themselves who paid the greatest attention to hydraulic projects.
Between 204 and 233 they constructed three large reservoirs, two trunk
canals, and six other important canals, and it was they who were
eventually victors due partly to their policy of developing military
agricultural colonies, partly to their policy of starving out their enemies
rather than fighting them, but also not a little to their efficient hydraulic
engineering.

During the destruction and devastation caused by the struggles of the
Three Kingdoms, it was natural that people should turn to an other-
worldly religion as a refuge. Buddhism was already there to fulfil
the need, but about the same time the Taoist philosophy (see pp. 110
onwards) united with the magical scientific elements in primitive North
Asian religion to produce an alternative and indigenous church. However, Buddhism became very popular, especially in later centuries, and

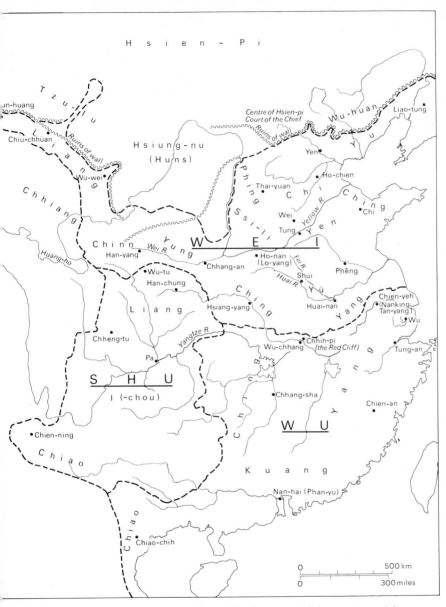

Fig. 10. China in the Three Kingdoms period (A.D. 220 to 265).
Based upon Herrmann, *Atlas of China*.

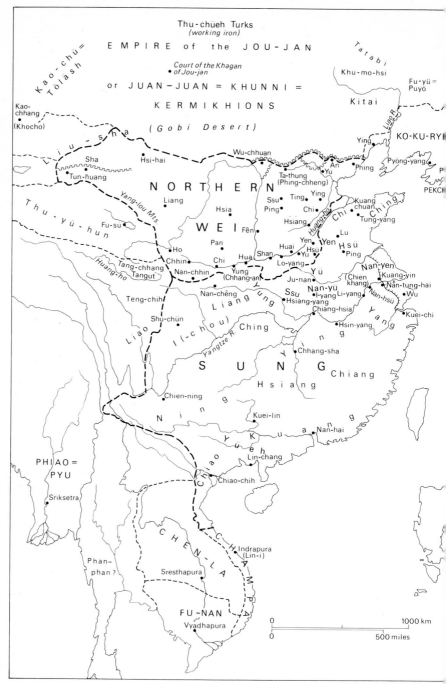

Fig. 11. China about A.D. 440. Based upon Herrmann, *Atlas of China*.

during the more settled times of the fourth and fifth centuries gave rise to a flowering of religious art, evidence of which is still to be seen in the famous carvings in the Yünkang caves in the east and the frescoes in the caves of the Thousand Buddhas at Tunhuang.

How imperishable the struggles of the Three Kingdoms became can be seen in the following story. Joseph Needham tells me that in 1943 he was sitting talking about the war with the peasant-farmers in a teahouse in Szechuan, and they said: 'You'll see, it will happen again as it did before, the North will win!' The Japanese ally Wang Ching-Wei was sitting in Nanking in the south like Sun Chhüan, ruler of Wu; the Generalissimo Chiang Kai-Shek corresponded to Chuko Liang in Chungking in Shu in the west; and far to the north Mao Tsê-Tung was the leader who represented Tshao Tshao of Wei. It was with some slight sense of shock that Needham and his companions realised the farmers were talking about the third century A.D. almost as if it was a few years ago.

The dynasty of Chin and its successors

Although the Wei gained control of the Three Kingdoms in 265, it was the Chin dynasty, founded the same year by Ssuma Yen, one of their generals, that ruled the newly unified state. Yet this did not mean peace for China, since almost at once the northern economic area came under pressure from the semi-barbarian peoples of the north, some of whom had already gained influence in China by becoming involved in its internal power struggles as allies. Within virtually the next half century, the Chin were driven south of the Yangtze and had to establish a new capital at Nanking. Yet in spite of their conquest, the northerners were never at peace among themselves, and for more than two centuries, between 304 and 535, no less than seventeen dynasties vied with one another for power. The Northern Wei was the longest surviving of these dynasties, eventually controlling the whole of the north, with the exception of Shantung in the north-east (Fig. 11). However, even though the north suffered continuously troubled times, it exerted its own influence on the invaders, who became increasingly sinified. Indeed, then, as at other times, Chinese culture displayed an astonishing integrative and absorptive power which no invader before the modern period could withstand.

As early as the latter half of the third century, at the beginning of the Chin, increasing foreign contacts stimulated the science of geography, especially under the great cartographer Phei Hsiu, and brought about the introduction of some new customs, among them the drinking of tea. All the same, the depopulation that had occurred due to the continual wars and skirmishes had its effect, and was probably the reason for the introduction of labour-saving devices like the wheelbarrow and the water-mill, and later, in the fourth and fifth centuries, for a developing military technology. One of the results of this technology was that military rather than administrative talents were sought after, with the result that those who did not have this flair began to turn more and more to speculation which, in its turn, led to a growth of theoretical science. Thus it was that in the fourth century the Taoists produced Pao Phu Tzu, one of their greatest naturalists and alchemists, while mathematics also flourished, and a new genre of writing, the gazetteers, appeared. Concerned ostensibly with local topography and records, they proved to be works of astonishing comprehensiveness, as the first, produced in 347, shows already clearly. Compiled by Chhang Chhü, and known as the *Hua Yang Kuo Chih* (Record of the Country South of Mount Hua), this gave details of the country south of Shensi and north Szechuan, described the building of the Shu capital, and then gave biographies of local notables. Yet these were only part of its contents; it also gave accounts of local monuments, described local customs, plants, birds and other animals, and provided information about the commodities available such as copper, iron, salt, honey, drugs, bamboo, and so on. Such books became widely popular and 6500 of them are known, although only comparatively few of these were written before the seventh century.

Political instability returned in the sixth century when the Northern Wei split into an eastern and western portion and then when both these were taken over in 550 by Chinese successor states, the Northern Chhi and Northern Chou. In the previous century there had been power struggles in the south also; the Chin had been replaced by the Liu Sung, which had itself given way to three other short-lived dynasties. Unification of both north and south did not come again until the 580s with Yang Chien and the Sui dynasty, which by 610 controlled the entire continent in a broad sweep taking in Annam and Formosa in the south to Tashkent and Sinkiang in Central Asia.

The dynasty of Sui

China had been divided for some 330 years, 60 under the Three Kingdoms and 270 under the 'Northern and Southern Empires'. The new unification, as might be expected, brought about improved connections between the northern and southern economic areas, the first Sui Emperor Wên Ti effecting some developments, and his successor, Yang Ti, overhauling the fragmentary water transport system between the Yellow River and the Yangtze, and building the main links in the first form of the Grand Canal. These new waterways passed right across the traditional battlefields between north and south, and formed a grandiose communication network that proved a blessing to later generations, yet they were only obtained at a great cost in human suffering. Some 5½ million people, including in some areas all commoners between the ages of 15 and 55, worked under the supervision of 50000 police, and those who could not or would not fulfil the demands made on them were punished by flogging and neck-weights. Every fifth family was also required to contribute one person to supply and prepare food and, in the end, the harsh conditions took their toll, and something of the order of two million men were said to have been 'lost'.

The Sui dynasty was too short-lived to have much effect on cultural matters. The cost of its public works and its extravagant military expeditions to Korea and Central Asia depleted the exchequer, and, coupled with other causes for complaint, brought public unrest that escalated into revolution when the emperor was surrounded by Turkic tribes at Yenmên and his power was obviously crumbling. Finally, in 617, power was seized by the official Li Yuan and his ambitious second son Li Shih-Min, who took the capital Chhang-an, and the next year proclaimed the Thang dynasty.

The dynasty of Thang

Building on the foundations laid by the Sui, the Thang emperors, whose reign was to last almost three hundred years, were able to enlarge China's boundaries and influence to an extent not reached since Han times, four centuries and more before. They repulsed attacks from Turkic tribes, taking the war into the nomadic lands where, thirty years later, the sovereignty of the Chinese 'khan' was finally accepted. They penetrated Tibet, whose king welcomed a Chinese wife and many civilising influences including technological developments like water mills and iron-chain suspension bridges, and by 660 they also ruled practically the whole of Manchuria and Korea, as well as Sinkiang. Maximum expansion occurred about 750, but after this there was a slow decline, triggered off by unfortunate diplomatic incidents in Tashkent, incidents that led in 751 to a clash between Chinese and Muslim armies at the Battle of the Talas River, and a resounding defeat for the Chinese. But the victory was a Pyrrhic one for although it resulted in the loss of Western Turkestan (part of modern Sinkiang), it also marked the end of Islam's expansion towards the east and was truly one of the decisive battles of world history.

The defeat of the Thang gave a signal to some countries within the vast Chinese Empire to seek their own independence: Mongolia was first, then the Thai principalities in the south-west, including Yunnan, rebelled and formed a separate dynasty, while in the north-east the Tartars set up strong bases in Manchuria, and south of them Korea absorbed the Chinese protectorates on its soil. What is more, good relationships with the Tibetans deteriorated until, in the end, they became such a menace to China and to the Muslim possessions in Central Asia that an Arab–Chinese alliance was formed between the Thang emperor and Hārūn al Rashīd, the famous Caliph immortalised in the *Arabian Nights*. After this a less unsettled period followed and the unification of the Thang continued for well over a century more.

Among the alternating periods of reception and rejection of foreigners that have characterised Chinese, as well as English history, the Thang was one when strangers were very welcome. The capital Chhang-an became an international meeting place; Arabs, Persians, Syrians came there to meet a diversity of other peoples, to discuss all manner of subjects with Chinese scholars in the elegant pavilions of the city in the Wei valley; and it became commonplace for wealthy Chinese to employ Central Asians as grooms and camel-drivers, Indians as jugglers, and Bactrians and Syrians as actors and singers. New foreign religions were imported: Zoroastrianism early in the sixth century, Christianity from Syria about 600, and Manichaeism from Persia at the close of the seventh century. As in Han times the Chinese themselves also journeyed far, the classic example being the Buddhist monk who went to India for 16 years, travelling the length and breadth of the subcontinent to gather religious writings. Buddhism, indeed, saw a period of great expansion and stimulated some of the finest artists of the period: the cave frescoes of Tunhuang are mostly from this time and reflect the general cosmopolitan attitude by showing monks and laity sometimes with brown or even red hair, blue or green eyes and even European features (Figs. 12 and 13). In the end, however, the Buddhist expansion, the proliferation of temples, the vast numbers who became monks or nuns, all began to seem like a state within the state challenging the accepted foundations of Chinese society. The administration became increasingly alarmed at this, and in 845 there was a vast purge; upwards of 4600 temples were destroyed and 40000 shrines abolished, some 260000 monks and nuns were secularised, 150000 slaves emancipated, and millions of hectares of arable land confiscated.

Yet if Buddhism appeared to present political dangers, one fruitful effect of its expansion was the stimulus it gave to the invention of printing. Buddhists required sacred pictures, repetitions of sacred names, and other similar items whose demand could best be met by block-printing. When this demand was coupled with the administra-

Fig. 12. General view of a cave-temple of Thang date (c. eighth century A.D.) at Chhien-Fo-Tung, Tunhuang. Predominant colours, pale red, pale green; with darker blue, green and brown in ceilings. Original photo.

Fig. 13. A military guardian deity (*lokapāla*) in one of the cave-temples of Thang date (*c*. eighth century A.D.) at Chhien-Fo-Tung, Tunhuang. On the back wall fresco, a scene of battle between good spirits and demons; on the side wall, the Western Paradise. Statues in reinforced plaster, about life size. Original photo.

tion's need for textbooks for those thousands of Chinese who were to sit civil service examinations, experiments in printing techniques were clearly called for, and these seem to have been started as early as the sixth century A.D.

It was during Thang times that Chinese laws were codified and a new criminal law developed, but the greatest glory of the period was the flowering of art and literature, and the foundation in 754 of the Imperial Academy (the Han-Lin Yuan or Forest of Pens). There was relatively little science: alchemy was much cultivated by Thang Taoists, cartography by Confucians, and some remarkable work was done by Buddhists, notably the astronomer and mathematician I-Hsing. He calculated the length of the year with considerable precision, and invented the first of all mechanical clock escapements. The general intellectual atmosphere however was humanistic rather than scientific. Technology showed some marked advances for apart from the first printing there was the manufacture of new ceramics, the Chinese at last producing a true porcelain which they handled with exquisite artistic taste. Also the millers were now in possession of the standard method of interconversion of linear and rotary motion, and in 659 the first official pharmacopoeia in any civilisation was produced.

Economically, the Thang dynasty tended to leave things as it found them. The state still leaned heavily on the peasant-farmers, and on merchants who did well enough in foreign trade but were continually frustrated by the absence of any banking system to help them. The administration, ever fearful of substantial accumulations of capital, was by nature opposed to giving any help in this direction; and indeed went so far as to reduce private mercantile growth by nationalising all tea exports. On the other hand the scholar-gentry thought it perfectly proper to amass wealth themselves and supplement their salaries by interest gained from lending their surplus money.

At the end of the ninth century and the beginning of the tenth, numerous small semi-independent states arose on the borders, and gradually the central government lost overall control. By 907 China reverted once more to separate areas of government, and the period of the Five Dynasties began.

The time of the Five Dynasties and the ten independent states

Severe fragmentation followed the collapse of Thang power, although why it was so extreme is not clear. After a thousand years of hydraulic engineering the country was covered with a network of waterways, and theoretically, at least, these should have helped perpetuate control by a central government. Perhaps the neglect of the network by the Thang administration after 750 had something to do with the case. But the development of new military techniques may also have been a factor, for the first references to the discovery and use of gunpowder take us back to 850 or 880, towards the beginning of the partition. Its first certain use in war was A.D. 919. Yet in spite of the great confusion that followed the overthrow of the Thang, some aspects of life and letters still managed to progress, and printing in particular made great strides.

The dynasty of Sung

Unification occurred once more in 960 when Chao Khuang-Yin came to power in a military coup and established the Sung dynasty. Often known as Emperor Thai Tsu – a posthumous title meaning Grand Ancestor – Chao lost no time in doing what he could to maintain the stability of his régime and insure against possible future military coups. To this end, he employed a brilliantly simple tactic. He wined and dined those generals who had been instrumental in bringing him to power, and then offered them each a large country estate and the means to run it if they would resign their military posts. So good was Chao's offer that all handed in their resignations the very next day.

In Manchuria and Mongolia, and the northern part of the North China plain, the old régime continued, but Chao ensured friendly relations by an annual consignment of gold and silk. A reversal of the traditional tribute to the Imperial Court, it was an effective measure nonetheless and set the pattern for systematic payments to barbarians. Peace reigned from 960 to 1126, but then, in spite of annual gifts, the Tartar armies of the Jurchen Chin swept down on the Sung capital Khaifêng, capturing the emperor and nearly all the government officials. Those who could, fled south, establishing a new government centred on Hangchow and known as the Southern Sung.

In spite of this disaster, the general pattern of Sung life seems to have remained unchanged. Hydraulic improvements continued, with water-conservancy projects numbering 496 compared with only 91 under the Thang, while the Sung also cultivated the arts and the sciences to a very high degree. Lyric poetry gave way to learned prose, religion to philosophical speculation, and if one follows up technological or scientific history, it is in the Sung that the focal point is always found (Fig. 14). Lock-gates and new surveying instruments, the caisson – a watertight chamber for use under water – and the transverse shear-wall for reinforcing the arches of bridges, were all introduced, and the period also saw the compilation of the greatest Chinese treatise on architecture. Ship-building advanced with the appearance of larger many-masted

51

Fig. 14. A Buddha or Boddhisattva in a cave-temple of Sung date (c. twelfth century A.D.) at Chhien-Fo-Tung, Tunhuang. Predominant colours, green, black, blue, white, gold. Original photo.

vessels, and navigation with the use of the magnetic compass. Chemical research progressed leading to the first printed scientific book in any civilisation. Gunpowder had found usage first in the form of grenades and bombs fired from trebuchets, then in a host of projectile weapons, especially rockets, the Sung–Chin wars providing their first major proving grounds. Chinese military technology never used the torsion catapults of the Greeks and Romans, but rather developed the trebuchet, a lever mounted on an upright post, which swung over to fling the projectile when the short arm was pulled down. They also developed powerful arcuballistae, i.e. sets of crossbow springs mounted on a carriage and shooting forth many arrows or bolts at one time.

If chemistry was applied to waging war, the biological sciences, on the other hand, were used for the benefit of man. Many famous physicians appeared during Sung times, old systems in pharmaceutics and acupuncture were improved and codified, and relatively new discoveries like variolation (a precursor of vaccination) were made known. Moreover in 1111 an Imperial Medical Encyclopaedia was compiled by twelve of the most eminent physicians of the day, while books on the pharmaceutical uses of plants reached an unprecedentedly high standard. The wood-block illustrations in these thirteenth-century pharmaceutical natural histories were superior to any to be found until their early sixteenth-century equivalents in Europe. The idea may even have been transmitted from the *Chiu Huang Pên Tshao* (Famine Herbal) of 1406. Specialised botanical and zoological monographs were also characteristic of the period, as too were the 'miscellaneous notes and records' which contained so many scientific observations.

It was during the Sung, in the eleventh century, that Shen Kua, perhaps the most interesting character in all Chinese scientific history, lived and worked. Born in 1030, he had the usual career of a scholar in government service – ambassador, military commander, director of public works and chancellor of the Imperial Academy – but wherever he went and whatever his official duties might be, Shen Kua noted down everything of scientific or technical interest. His *Mêng Chhi Pi Than* (Dream Pool Essays) of 1086, or thereabouts, is one of the first books to describe the magnetic compass, and contains much astronomy and mathematics, instructions on making relief maps, descriptions of metallurgical processes, notes on fossils, as well as a host of biological observations. Science in fact occupies more than half the book.

Interest in mathematics was not confined to Shen Kua, for the Sung produced some of the greatest Chinese mathematicians of any period in history. Chhin Chiu-Shao, Li Yeh and Yang Hui developed algebra to the highest level in the world at that time. And humanistic studies did not lag far behind the achievements of the mathematicians; in 983 a chronological encyclopaedia appeared giving systematically arranged quotations from a host of ancient and mediaeval authors, a geographical encyclopaedia followed soon afterwards, and in 1084 Ssuma Kuang produced the first complete history of China down to his time. These were compendious works, a tribute not only to Sung scholarship but also to a satisfying intellectual environment, one which also encouraged the growth of Neo-Confucianism. This was a kind of scientific human- ism and we shall come back to it later.

Besides this upsurge of cultural activity, the Sung period also pro- duced Wang An-Shih, the second of the two great reformers in Chinese history. Unlike Wang Mang he never attained or desired the Emperor's Imperial Yellow, but like him he took a lively interest in science and technology, and studied botany, medicine, agriculture and weaving. After 1069, when he became a minister, he initiated reforms that have become famous ever since but aroused a storm of opposition at the time. Mainly financial, they began with a reform of the Treasury, reducing the embezzlement that went on and rationalising the administration so effectively that he was able to save some forty per cent of the national budget. Wang also abolished the system of transporting grain to the capital, establishing government warehouses in all large cities from which the grain could be sold directly instead. The taxes, which he based on new land surveys, could therefore be transmitted to the capital in cash. Government advances were instituted for farmers on the security of growing crops and at cheaper than market interest rates, while money payments were allowed in lieu of forced labour. Production of luxury goods was restricted and any hoarding of commodities was heavily taxed. In addition to these measures, every ten families was grouped into one unit, and all members of that unit were responsible for the misdeeds of any one member. These units were also used as the basis for army conscription, although large landowners had to furnish horses instead of men.

Opposition was vehement. The agricultural population, glad enough at the repeal of forced labour, was nevertheless totally opposed to the conscription scheme and to the tyrannies that the group responsibility method brought in its train: the gentry objected to the abolition of peasant labour, state officials to the strict accounting that prevented them from what had become a legitimate, if modest, means of enrichment, and both officials and gentry had a profound distrust of any reliance on paper money. In brief, although the reforms showed a surprising degree of original thought, they were too radical a departure from traditional feudal bureaucracy. Wang An-Shih died in retirement and apparent defeat in 1086, although his policies were not forgotten and some came gradually into effect under others.

The Yuan (Mongol) dynasty

In the thirteenth century there occurred the greatest clash in Asian history between the nomadic culture of the steppe and the civilisation based on intensive agriculture. In 1204 Chinghiz was proclaimed Khan of all the nomadic Mongols and he set about a policy of expansion, first attacking the Jurchen Chin Tartars and taking Peking in 1215, occupying the Hsi-Hsia Kingdom twelve years later, and then conquering Khai-fêng in 1233. The Mongols now settled down to overcome the Sung and for the next forty-five years struggled against the best equipped and subtlest of their foes, until in 1279 the last Sung prince was killed in a sea battle and they could at last claim to be masters of the whole of China.

The Yuan dynasty lasted for just over a century, although Chinghiz never ruled China himself, having died in the Hsi-Hsia campaign. When they eventually occupied the Sung country they were astonished at the agricultural wealth that they found. To begin with they thought of massacring the population and turning the land over to pasture, but Yehlü Chhu-Tsai, an adviser descended from the royal house of Liao, persuaded the Mongols that they would do better to receive the dues already coming from a well-established taxation system, and with Kuo Shou-Ching, a Chinese, set about administering the country under its new masters. Both men of science, they found time, in spite of their heavy programmes of work, to establish at Peking one of the most important astronomical observatories of the age.

Under the Yuan, China became better known in Europe than at any time in its history up to the present century. This was due to the vast territory which the Mongols now controlled, an area that included the country north of the Himalayas from well east of Peking and westwards as far as Budapest, and in the south from Canton westwards to Basra. With the entire domain under one unified control, the roads across Central Asia were safer than ever before or since, and the Khan's court became filled with Europeans or Muslims who had some skill or craft to offer. Indeed, although Chinese were recruited into the administration, they were not trusted with the highest posts which went to foreigners if no Mongol was suited to fill them: thus it was that Marco Polo spent some sixteen to seventeen years as a court official. Within China too, communications were improved by the Yuan; roads were laid with post-stations along them, and the canal system was extended. Expeditions were sent to determine such matters as the source of the Yellow River, and geography flourished, Chu Ssu-Pên producing his great atlas, the *Yü Thu*, between 1311 and 1320.

In Yuan times Christians made a second attempt at penetrating China, but the Franciscan friars achieved no more success than the Nestorian clergy six centuries before. Taoism also suffered; its books were burnt and the movement was driven underground, although this only encouraged it to develop into a national cult associated with the struggle against foreign domination. Gradually, as the years passed, the Chinese scholar-gentry infiltrated back into the administration, Confucianism gained much ground, and by the middle of the fourteenth century the days of the Yuan dynasty were seen to be numbered. Having failed to organise the bureaucracy adequately, it ran into financial difficulties, and secret societies dedicated to the expulsion of the Mongols became increasingly active. By 1356 Nanking was captured by a nationalist movement, and a decade later a Ming army took Peking. Finally, in 1382, the last Mongol strongholds were overthrown with the conquest of Yunnan province.

The Ming, and Chhing (Manchu), dynasties

The Ming established their capital at Nanking in the east-central economic area. They promulgated new laws, undertook fresh irrigation-works, and waged war on the Mongols, sacking their capital at Karakorum and pursuing the fugitives as far north as the Yablonovy mountains, further than any Chinese army had ever gone before. Manchuria was annexed, but for a time after the death of Hung-Wu, the first Ming emperor, China itself suffered a civil war while the emperor's successors struggled for control. Eventually, in 1403, peace returned, the capital was shifted to Peking, and under Chu Ti, the third Ming emperor, the country entered the Yung-Lo period.

The Central Asian territories may have contracted, but this was the time when the Chinese embarked upon their greatest period of maritime exploration. In 1405 the eunuch admiral Chêng Ho left with a fleet of sixty-three ocean-going junks, visited many parts of the South Seas, and brought back with him the kings of Palembang and Sri Lanka to do homage at the Imperial Court. During the next thirty years, seven similar expeditions set forth, and all returned with geographical information and produce. For the first time animals such as ostriches, zebras and giraffes came to be seen in China. The motive behind all these expeditions is not clear, although official annals say that they were aimed at seeking out the emperor's predecessor who had fled at the end of the civil war and had, so it transpired later, gone into hiding as a Buddhist priest. More likely the expeditions were for state trading, new drugs, minerals and natural curiosities. At all events, they stopped as suddenly as they had begun, and the command of the Indian Ocean was left to the Arabs and the Portuguese. Soon Japanese pirates preyed on the

Chinese coast, and Annam, which had been annexed by Chu Ti, became independent once again. From 1514, some eighty years after Chêng Ho's first expedition, Europeans began frequently to approach the coast, the English first arriving in 1637. Russian attempts to make contact by way of Siberia were unsuccessful, but the Spanish, who occupied the Philippines in 1565, did better; they bought books, established some trade, and introduced the Mexican silver dollar into Chinese commerce.

Domestically China restored her native culture during Ming times. Defences, paved highways, rock-gardens, bridges, temples and shrines, tombs and memorial arches, were all built in great profusion, and the walls of 500 cities underwent complete reconstruction. The bureaucracy rose again, and in 1469 there were some 80000 military officials, with over 100000 civil servants. However, the bureaucracy was soon disrupted by an internal power struggle between the eunuchs and the Confucian scholars, the eunuchs eventually getting the upper hand. Excluded from the bureaucracy, the scholars formed themselves into organisations that were partly academies and partly political parties. These engaged in a vast amount of academic work that characterised the fifteenth and sixteenth centuries, when there was a flourishing encyclopaedic movement. Its most impressive result was the vast *Yung-Lo Ta Tien*, a compendium of more than 11000 chapters commissioned in 1403, and so well organised that it took its 2000 scholars only four years to complete. The original contained copies of rare books, but it was considered too large to print and only two further copies were made. Unfortunately the main collection was destroyed in 1901 during the Boxer Rebellion, but some 370 volumes still survive in libraries all over the world. In philosophy, the age was dominated by Wang Yang-Ming, who moved away from scientific humanism to a rather anti-scientific idealism, although, on the other hand, a scientific approach to phonetics flourished at the same time.

One must beware, however, of gaining the impression that there was no real science during the Ming period. Geography flourished, while imperial princes as well as commoners took an interest in botany; a great botanic garden was maintained close to Khaifêng, and in addition a 'Natural History for Famine Times' appeared in 1406 with fine plant illustrations. But undoubtedly the greatest scientific achievement of the Ming was the *Pên Tshao Kang Mu* (The Great Pharmacopoeia) of Li Shih-Chen, which appeared in 1596. It described about 1000 plants and 1000 animals in exhaustive detail, classifying them into 62 divisions according to their ecological character, while an appended work added more than 8000 prescriptions. In this book Li Shih-Chen also gives interesting discussions on distillation and its history, smallpox inoculation, and the use of mercury, iodine, kaolin and other substances in therapeutics. Two important technological works also appeared, one describing every kind of manufacturing process, and the other dealing with all aspects of military technology.

By the beginning of the seventeenth century, the position of the Ming Government was deteriorating; taxation was high and many abuses had grown up. In 1636 Manchuria was lost and eight years later a rebel leader was treacherously admitted into the capital, whereupon the last Ming emperor committed suicide. Manchu aid was sought to help in restoring order, but once called in they refused to leave, and from then on the Chhing dynasty ruled.

This is the point at which we may stop our historical résumé, for in 1582 the Italian Jesuit Matteo Ricci reached Macao, and in 1601 went on to Peking where he died nine years later. An extraordinary linguist, scientist, geographer and mathematician, he assimilated himself and his Jesuit colleagues into Chinese society. Accepted at court, Ricci was soon busy reforming the calendar and arousing widespread interest in science and technology. With a few learned converts, especially the agriculturist and official Hsü Kuang-Chhi, he made translations of Greek books on mathematics, astronomy, and hydraulics, and encouraged the compilation of works by other scholars; Hsü Kuang-Chhi for example wrote a complete treatise on agriculture. Thus it was that Greek and European Renaissance science reached China: during the eighteenth and nineteenth centuries it slowly fused with Chinese science and with universal world science until now it is no longer possible to detect any particular style in the subsequent contributions of Chinese thinkers and observers. This is why we reach the limits of this review of the linked development of science and civilisation in China.

6

The travelling of science between
China and Europe

The originality of Chinese culture

Although there is a copious literature on the contacts between
China and Europe, the facts are known to comparatively few. For
example, a belief that much characteristic development of Chinese
thought and practice derived from the West still lingers on; among other
things early Chinese astronomy, the evaluation of *pi* (π), or hydraulic
machinery, have all been said to owe their origin to the West. This we
now know to be untrue.

It would be a mistake, moreover, to think that every development
had only one source of origin; we cannot rule out the possibility of
completely independent and parallel lines of thought occurring in widely
separated parts of the world, especially where scientific discoveries are
concerned. For instance, it now seems likely that the biological idea of
a 'ladder of souls' arrived independently in both East and West. In the
West it was Aristotle who propounded the doctrine that plants had a
vegetative soul, animals possessed vegetative and sensitive souls, while
man had both and a rational soul as well. This was in the fourth century
B.C., yet less than a century later Hsün Chhing was teaching the same
concept in China. Clearly Aristotle's theory could conceivably have
travelled all the way from West to East, but it would have had to do
so very quickly, and at a time when conditions of travel made such a
transmission most unlikely. What is more probable is to suppose that
the idea, which essentially reflects an appreciation that some living things
are more complex than others, arose independently in both places.

There is, however, no doubt that China was in the circuit as far as
the general diffusion of knowledge was concerned; indeed, there was
probably far more exchange between China and its western and southern
neighbours than has often been supposed. Moreover, Chinese inventions
and discoveries passed in a continuous flood from East to West for twenty
centuries before the scientific revolution, and the more technological

they were the easier was their travel. The distinctively scientific ideas of the Chinese tended to stay at home, probably because their natural philosophy was not easily integrated with, or comprehensible to, the Westerners and their conceptions. Yet all these interchanges never affected the essential style of Chinese thought and culture patterns; which maintained a remarkable independence in spite of any diffusion. Contacts there were, interchanges of ideas and techniques certainly occurred, but never in such profusion as to affect the characteristic style of China's civilisation, and hence of her science. In the light of this it is no exaggeration, then, to speak of the 'isolation' of China, or, alternatively, of the rest of the world from her; culturally that was something very real.

Many examples of Chinese influence on Europe in the field of applied science will come to light in what follows, yet even when we have good reason to believe in a transmission from China to the West we generally know very little of the means by which it took place. But there is a principle operating here by which Joseph Needham and his collaborators set much store, namely that where there is doubt, the burden of proof lies upon those who wish to maintain completely independent invention or discovery; and the longer the period elapsing between the successive appearances of such a new device or development in two or more of the cultures concerned, the heavier that burden generally is. Techniques were of course frequently improved upon by their recipients, as the centuries passed, but who knows what capillaries of transmission were not there to connect the original discoverer or inventor with his successors in quite different civilisations of the Old World?

The continuity of Chinese with Western civilisation

If China retained her individuality this was not because diffusion contacts with the West were a late development; on the contrary they began at least as early as the Bronze Age, that is before the Shang dynasty (1500 B.C.) and well into the Chou. At these times it may well have been that this occurred because less rigid national barriers separated different countries, but whatever the reason there is no doubt at all that interchanges went on; much evidence is available to prove it. For example, similar forms of bronze swords appeared in China and the West (Fig. 15), ceremonial axes showed affinities in style (Fig. 16), so did harness pieces and other bronze accoutrements. Again, the 'bird chariot' – a bronze or pottery image of a bird on three wheels – is to be found not only in China, but also in Egypt, and some central and western European sites, although whether this was a toy or a religious cult-object we do not know. Then there were fabulous creatures and

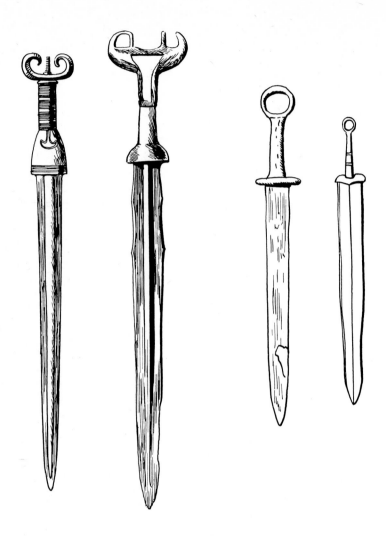

Fig. 15. Two-edged bronze-age swords with antenna-pommels and ring-pommels, showing the technological continuity between Shang and Chou China and Europe at the time of the Hallstatt culture. Left to right: Denmark, China, the Kuban (Russia), China. Reproduced from O. Janse, 'Notes sur quelques Epées Anciennes Trouvées en Chine', *Bulletin of the Museum of Far Eastern Antiquities*, Stockholm, 1930, **2**, by permission.

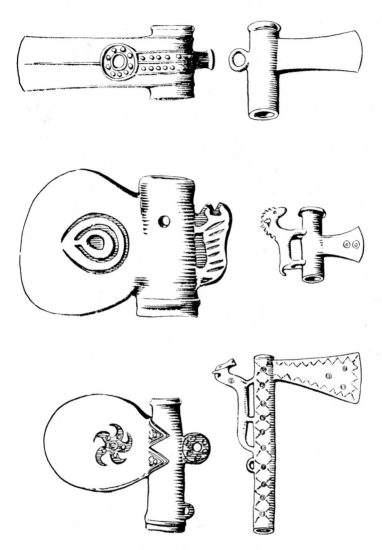

Fig. 16. Bronze-age ceremonial axes; a non-functional blade opposite an animal in full relief, or a ring. On the left three Chinese examples, on the right three from the Hallstatt culture. Reproduced from O. Janse, 'Quelques Antiquités Chinoises d'un caractère Hallstattien', *Bulletin of the Museum of Far Eastern Antiquities*, Stockholm, 1930, **2**, 177, by permission.

伏戲倉精初造王業畫卦結繩以理海內

Fig. 17. Wu Liang tomb-shrine relief (second century A.D.) showing the deified culture-heroes Fu-Hsi and his sister-consort Nü-Kua as unipeds, with the carpenter's square and *quipu* (itself personified) as symbols of construction and order. The inscription says: 'Dragon-bodied Fu-Hsi first established kingly rule, drew the [eight] trigrams, [devised] the knotted cords [*quipu*, for reckoning], in order to govern [all] within the [four] seas.'

mystic plants that the various civilisations held in common, like the divine beings with mermaid tails (Fig. 17) which appear in China and in Gaul, and the magic of mugwort (*Artemisia argyi*), which, in spite of its unimpressive appearance (Fig. 18), was used as a constituent of incense and a powerful charm against demons in Mexico and in ancient and mediaeval Europe, as well as China. Myths and religious rites also appear to have been transmitted to some extent one way or the other.

Trade-routes between China and the West

Additional evidence of contact may be found by examining the Western names for China, the best known of which were probably Seres, Sina, and Cathay. Seres comes from the Chinese *ssu* (絲) silk, and was

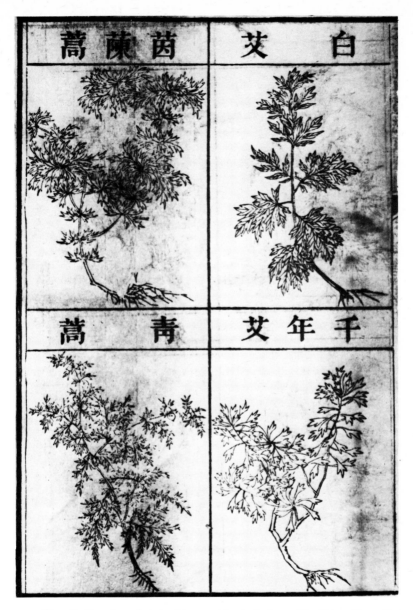

茵陳蒿   白艾

青蒿   千年艾

Fig. 18. Drawing of *Artemisia argyi* (*pai ai*), from *Pên Tshao Kang Mu* (A.D. 1596), top right-hand picture. Li Shih-Chen's great pharmacopoeia recounts the many uses of this plant and its relatives in Chinese materia medica. They were well known as anthelminthics, and indeed this genus is the source of santonin, an alkaloid still most valuable against nematode parasites. The dried and powdered leaves of *Artemisia* formed the tindery basis of the moxa cautery, so much used in old Chinese medicine.

transmitted to Europe as *ser* by the Greeks; it had obvious connotations with the silk trade from perhaps as early as 220 B.C. *Sina* is Latin and from it we derive our name China, but it seems to have come not from Rome but in the second century B.C. by way of India: it is a corruption of a Sanskrit form of the Chinese dynasty name Chhin. Later dynasties gave rise to other names, and the tenth-century tribal name of Chhi-tan Liao was transformed to the Russian *Khitai*; this became the European *Cathay*. In the Middle Ages there was much uncertainty about whether China and Cathay referred to the same country, just as there had been in classical times over Sina and the Seres, but in each case the doubt seems to have arisen because one name had arrived by overland trans-mission, the other by sea.

Classical western references to China, however, go back long before the report of the Seres. As early as the fifth century B.C. Herodotus refers to what appear to be the Chinese in his detailed account of the Central Asian Scythians and their contacts. In the course of his descriptions, based on travellers' reports, Herodotus mentions the Hyperboreans, a tribe which some modern research has identified with Chinese dwelling in Kuanchung and the lower Yellow River regions. This interpretation may seem somewhat strange until it is realised that Herodotus' world map would indeed show people in China as living 'beyond the north wind' and enjoying a paradisiacal mild climate on the other side of the terrible winter of Central Asia.

The next Greek scholar to concern himself with the Chinese was the astronomer and geographer Ptolemy of Alexandria, who, in the second century A.D., provided a tolerable account of the whole country between the Caspian Sea and China. Much of his detailed information came from the agents of the silk merchant Maës Titianus who travelled the Old Silk Road, and this growth in Greek knowledge about China between the fifth century B.C. and the second century A.D. reflects what we already know from the Chinese side. When in the second century B.C. Chang Chhien went on his protracted embassy (see p. 37) he found some sort of trade route already in existence between India and the west of China, running from Szechuan southwards by way of Yunnan and either Burma or Assam. Such a route to India, coupled with other routes from India to the Middle East, would explain how he could bring back information on countries as far off as Parthia and Syria. It was Chang Chhien's journey that paved the way for the Old Silk Road, that Titianus' agents were to use, a road that did more, however, than act as a route for the export of Chinese silk to the West. The Old Silk Road was also used for imports into China, especially plants like the grape vine as well as alfalfa, chives, coriander, cucumbers, figs, safflower, pomegranates, sesame and walnuts, half of which have the character *hu* (胡) in their names, including their origin in Central Asia or Persia. The traffic in plants was not, of course, one way: from China westwards went oranges and, in due course, pears and peaches, which reached India by the second century A.D. Many centuries later China was also to provide an altogether surprising proportion of the cultivated flowers now to be found in Western gardens: roses, peonies, azaleas, camellias and chrysanthemums.

Overland routes were only one means of travelling from China to the West; there were also sea routes by way of India, routes that involved Greek and Roman contacts not with Central Asia but with India itself. The earliest information the Greeks had of India came at the close of the fifth century B.C. from the physician Ctesias of Cnidus, who heard of the country while at the Persian Court. Nevertheless, Ctesias' accounts were rather too imaginative, containing, for instance, references to 'fountains of gold' instead of abundance of the molten metal, and 'tree garments' in place of cotton-growing. Not so the reports by Megasthenes, Seleucid ambassador between 302 and 308 B.C. to Chandragupta Maurya, the first ruler to unify the greater part of the Indian subcontinent. Although long disbelieved, his accounts were much more accurate, including a description of the opinions of the Brahmin philosophers and the Buddhist monks, but even he believed some stories like the one that alluvial gold grains were obtained by the activities of 'gold-digging ants'. However, one of the most interesting and valuable items of evidence of contact between India and the West was an anonymous book *The Periplus of the Erythraean Sea* (i.e. the Indian Ocean). Written about A.D. 70 by a Graeco-Egyptian seafaring merchant, it described not only the ports but also the commodities bought and sold there, and did so with a certainty born of intimate personal experience. The wide extent of at least commercial contact by this time is clearly evident.

Maritime trade developed slowly but steadily from the first to the middle of the third centuries A.D., with ships Roman in name though really Graeco-Egyptian reaching all parts of India, and towards the end of this period, even penetrating as far as Kattigara, which was either Indo-China, or perhaps the coast of south China itself. Syrian and Graeco-Egyptian 'colonies' or 'factories' may have been established at Canton and Hangchow. Indian and Singhalese ships also plied the same routes, and a few Roman trading settlements were established in India, but it was not until after the third century that long-distance Chinese navigators appear on the scene. By the eighth century the picture had changed yet again, the Arabs were masters of these seas, and during the ninth century maintained direct contact with the Chinese, frequently visiting the south China coast and themselves establishing 'colonies' or 'factories' at Canton and Hangchow. But Arab supremacy did not last for ever, and after the end of the twelfth-century Arab navigation, which had even extended to the Pacific, gave place to that of the Chinese, who in the fifteenth century had a short period of maritime supremacy under the Ming. This brought Chinese junks to Borneo, the Philippines, Sri Lanka, Malabar, and even Zanzibar in East Africa until, in the sixteenth century, the modern era opened with the Portuguese explorations.

The Old Silk Road

While the growth of maritime trade was occurring, overland travel was also being developed. By 106 B.C., a couple of decades after Chang Chhien's mission, a trans-Asian silk trade had become regularised, and a number of routes were in use (see Fig. 19), though by A.D. 400 increasing desiccation of the surrounding land led to the abandonment of the city of Loulan and the closure of the particular road running through it. The roads passed through many separate countries and a number of cities, so there were many middlemen who had a hand in the silk trade and took their cut as it travelled on its way to the West. It was the attempt to avoid at least some of them that led to the establishment of so many alternative routes. Apart from silk and the plants mentioned previously, commodities like lacquer boxes and vessels travelled westwards over these trade routes, together with ivory carvings, spices and that Seric iron (probably steel) mentioned by the Roman encyclopaedist Pliny the Elder. Rhubarb seems to have had a special route of its own. From West to East went glass, wool and linen textiles and artificial gems, but not much else; the Romans had an adverse trade balance made good by bullion which probably never reached China because absorbed by the middlemen on the way.

When the Roman Empire broke up what can best be described as a series of shock waves travelled throughout Asia. Yet in spite of such far-reaching political reactions, there was no interference with the chief land route, although the sea route suffered some change, due primarily to the growth of the kingdom of Axum, with a port Adulis at the lower end of the Red Sea which became a centre for the ivory trade and was able to assume as vital a middleman status as any Persian centre. However, with the founding of Byzantium early in the fourth century, new land routes came to be developed from Merv (close to present Marv in the Turkmeniya Republic of the U.S.S.R.), running through Armenia and bypassing Mesopotamia; indeed, silk became one of the foundations of Byzantine prosperity. Nevertheless the latter were in a

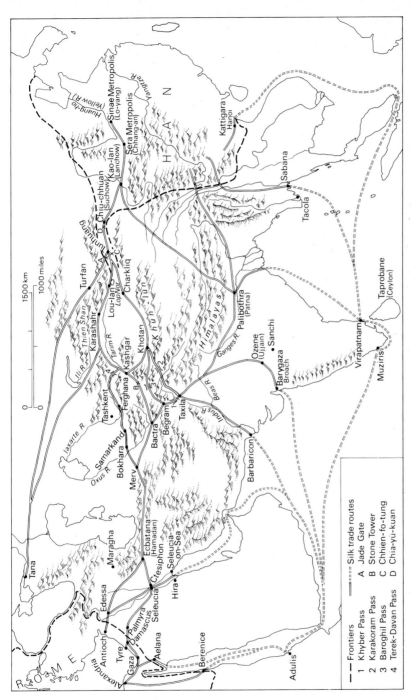

Fig. 19. Routes of trade between China and the West from the first and second centuries A.D. onwards. Based upon G. F. Hudson, *Europe and China*, Arnold, London, 1931, by permission.

difficult position since they were forced into wars with the Persians, the very people on whom they depended for their silk supplies.

Suddenly, in 552, a decade after Justinian had nationalised the Byzantine silk trade, the whole difficult situation was altered by the introduction of the silkworm into Europe. There are different accounts of how this happened: the Byzantine writer and senator Procopius said that certain Indian monks offered to bring it from 'Serinda', and the chronicler Theophanes the Confessor claimed that some Persian monks smuggled the eggs of the moth out of China in a hollow cane. Both accounts can be reconciled, however, if the monks were Persian Nestorians who had first been to India and had brought the eggs from Cambodia or Sinkiang. Yet this is only part of the story. To transplant a silk industry needed more than a silkworm's eggs: it needed the presence in the West of people knowledgeable in the whole technology of silkworm cultivation, and of how this came about we are told nothing.

The land routes from China continued to hold even during the sixth-century Turkish domination of Central Asia, while during the seventh and eighth they enabled Byzantium to establish embassies in China; but the Battle of the Talas River in 751 led to a closing of the route that passed close by. What was lost on land, though, was compensated for by Arab mercantile exploration eastwards and expansion of the maritime routes. Moreover, from the seventh to tenth centuries, during the period of the Thang, Persian merchants were always to be found in China, having travelled presumably by sea if their land route happened to be closed. Persian embassies had been accepted long before this, the first having come in A.D. 456 with gems, carpets and aromatics, but by Thang times the Persians themselves were popularly regarded in China as magicians, alchemists and wealthy dealers in precious stones of magical power. Their centre was the western market at Chhang-an where, it seems, they often paid high prices for jewels or minerals little valued by the Chinese.

During the eleventh and twelfth centuries the land routes had a reduced significance, but under the Mongols they revived until they reached a hitherto undreamed-of importance. This was because the extension of Mongol power as far westwards as Baghdad and Budapest meant a safe road from China right into the western world, so secure that a fourteenth-century merchants' handbook recorded 'The road which you travel from Tana to Cathay is perfectly safe, whether by day or by night.' Tana was at the mouth of the Don.

In the thirteenth and fourteenth centuries, the Mongol Khans collected European technicians at their courts, though there is little evidence that this resulted in the transmission of mechanical arts or natural principles. On the other hand, during the fourteenth and fifteenth centuries there was a trade in Tartar (Mongol and Chinese) slaves with the West. Italian homes had Tartar domestic servants; between 1366 and 1397, for instance, no less than 259 Tartars, mostly young women, were sold at the slave market in Florence. The influx, which had begun in 1328 when Marco Polo's own servant Peter the Tartar was granted Venetian citizenship, and ended in 1453 with the fall of Byzantium, gave rise no doubt to much inter-racial mixing and may have led to the transmission of 'know-how' in several branches of technology, especially textiles.

Chinese–Western cultural and scientific contacts

The trade routes between East and West led to the exchange of diplomatic missions between China and various Western countries, although the balance was unequal, more reports going to China from the West than came from China westwards. Indeed the Chinese seem to have been less adventurous than the Westerners, less interested in foreign cultures. While the Han, for instance, were always ready to welcome visitors, they were less willing to brave the rigours of return journeys overland on a route that they believed to be hazardous and on which one might suffer the inconveniences of mountain-sickness. Even those who did go seem often to have been ready for an excuse to terminate a journey. A typical example, from the first century A.D., was the ambassador Kan Ying. On his way to the Roman Syrians, he allowed himself to be persuaded by Parthian mariners that the dangers of taking the sea voyage from Babylonia were too serious for him to go further. Doubtless the Parthians exaggerated the perils in order to avoid direct contact between Romans and Chinese, since this would have been to their disadvantage as middlemen; indeed it seems this was a general ploy, and one that was so successful that no Chinese agents ever did reach Rome itself.

Of the many visitors to China, not all were attached to embassies or to parties of merchants; there was an influx from Syria of acrobats and jugglers, and although it may seem unlikely on the face of it that such people should have anything to do with the transmission of technology, the case is probably otherwise. In those times much of the inventive effort of Western engineers like Ctesibios (second century B.C.), Heron (first century A.D.), and their Chinese counterparts Ting Huan and Ma Chün, was concerned with conjuring illusions, mechanical toys for palace entertainments, machinery for stage plays, and the like. The methods they used often contained technological innovations, and we know for certain that it was just such kinds of 'marvellous' entertainment

Fig. 20. Illustration of a striking water-clock from a MS of al-Jazari's
treatise on mechanical contrivances (A.D. 1206). At the top, the signs
of the Zodiac exhibited, then figures successively appearing and
lamps successively illuminated, below that, the golden balls dropped
into brazen cups from the beaks of brazen falcons to strike the chime;
lastly an automaton orchestra of five musicians. Such Arab striking
water-clocks (derived from Byzantine origins) were described by
Chinese historians in the tenth century A.D., and the Chinese had
themselves been constructing them since the seventh century A.D.
Reproduced from K. A. C. Cresswell, 'Dr F. R. Martin's Treatise on
Automata', *Yearbook of Oriental Art and Culture*, ed. A. Waley, Benn,
London, 1925, by permission.

that appeared in China with the first Parthian missions. Indeed, five centuries later, we still find descriptions of visiting magicians and conjurers. Later still, from Thang times onwards, official records also note the appearance of other marvels, especially the arrival of clepsydrae (water-clocks) from the West, float clocks of a type familiar to the Chinese but with novel and complex striking mechanisms, many of which allowed golden balls to drop one by one into a receptacle to mark the passing hours, a design that seems to have been a continuation of an original Byzantine tradition (Fig. 20).

Among the other wonders that made their appearance in China, there was the 'night-shining jewel' from Syria. Precisely what this can have been is uncertain, although it was most likely a stone cut from the mineral fluorspar, of which many varieties glow on being heated or scratched in a dim light. There were also coral and pearls imported from the Red Sea, and amber that came either from the Baltic or from Syria. Yet the Chinese soon realised the artificiality of many of the 'gems' that accompanied the genuine precious stones. Of course, the appearance of such 'paste' jewellery should be no cause for wonder, since much of the wealth and opulence of the time, the kind of fabulous richness in the accounts of East Roman palaces, for instance, was very probably 'imitative', with gilt copper and coloured glass everywhere; Syria, situated among the gem-producing districts, was ideally placed for making copies of this kind. In view of the interest in artificial imitations in early European proto-chemistry, the Chinese ability to sort the genuine from the false is quite significant from the point of view of the history of science.

Besides records of imports of a technological kind, there is also, in the Thang dynasty, one isolated case of a quite different kind: interest in a Western medical report. The subject was trepanning (cutting a hole in the skull) by Roman Syrians who had 'clever physicians who by opening the brain and extracting worms, can cure *mu-shêng* (a kind of blindness)'. Trepanning itself had been known all over the ancient world since Palaeolithic times, but the Chinese report is what is interesting here. Similar operations for intra-cranial cysts or tumours seem to have been known in China about the same time. The Western belief in a universal antidote to poisons which could have anything up to 600 ingredients also found its way to China, but such a panacea was but a myth, as the Chinese quickly recognised.

Chinese–Indian cultural and scientific contacts

There is early evidence for contact between China and India in Chang Chhien's description of products from Szechuan that he found had reached Bactria, though the greatest period of contact began in the middle of the fourth century A.D. with the Buddhists. In 386 the monk Kumārajīva went to China propagating Mahāyānist Buddhism (interpreting the earthly Buddha as a manifestation of a transcendent celestial Buddha), but he was only the first of a long succession. Indeed, this was the time when many Chinese Buddhists made pilgrimages to India, visiting the courts of foreign rulers where they described life at home, then writing up their travels when they returned. This was a genuine two-way intercourse. Seventh- and eighth-century texts describing Brahmin astronomy and mathematics, and giving what may be the earliest passage on mineral acids, bear evidence of this contact.

Yet in spite of these relationships, definite proof of Indian influence on Chinese science is hard to find; only the effect of Chinese mathematics on Hindu mathematics is unmistakable. The Chinese acceptance of concepts like that of the 'lunar mansions' in astronomy, and subtle vapour or 'pneuma' in medicine, seems to have come rather from Mesopotamia, travelling outwards by direct routes to the Greeks and China as well as to India. So far as Indian technology is concerned, the situation is similar: it affected China but little, apart from the noria or water-raising wheel with rim-buckets, and the textile machinery for cotton which came later.

On the whole, then, Chinese contact with India had little scientific or technological effect, though the great achievements of Indian grammar and linguistics undoubtedly influenced fifth- and sixth-century Chinese philological study, and Indian painting and music also made a definte impression on the Chinese.

Chinese–Arab cultural and scientific contacts

By 625, only three years after the flight of the Prophet to Medina, the new expansionist power of Islam that was to affect East–West relations so profoundly, had conquered Persia and brought Arab dominion direct to the frontiers of Chinese influence. Yet the first Islam–Chinese contacts have gathered so many legends in the telling that it is now difficult to disentangle fact from fiction; certainly at one time there were Islamic hopes of conquering China, but in the event no Arab army ever set foot there. Intellectual contacts there were in plenty, however, from the seventh century onwards, and after Harūn-al-Rashīd's mission in 798 for concerted Arab–Chinese action against the Tibetans, trade relations and all that accompanied them flourished most effectively and continued, it seems, uninterruptedly for the next three hundred years, until well into the eleventh century.

During the late thirteenth century, after the Mongol conquests, a new dimension arose in East–West unity, and personal contact between

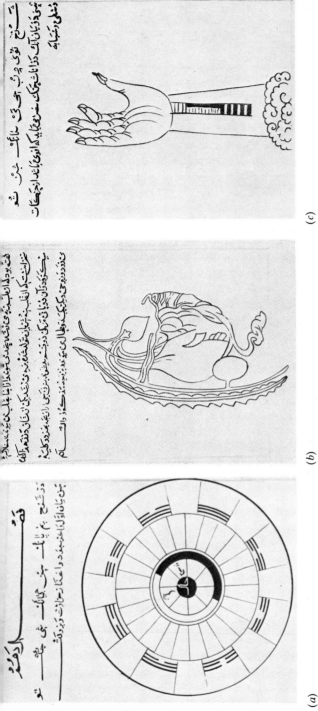

(a)

(b)

(c)

Fig. 21. Three medical drawings from the *Tanksuq-nāmah-i-Ilkhān
dar funūn-i 'ulūm-i Khitāi* (Treasures of the Ilkhan on the Sciences of
Cathay), prepared by Rashīd al-Dīn al-Hamdāni about A.D. 1313. In
(*a*) the aegis of the Eight *Kua* over the day and night of the patient is
related to the rise and fall of fever; (*b*) shows an anatomical sketch of
the viscera (heart, diaphragm, liver and kidney can be made out);
and (*c*) is a pulse-lore diagram. Though the text is in Persian, the
Chinese origin of the illustrations is quite clear. Reproduced from A.
Süheyl Ünver, *Tanksuknamei Ilhan der Fünenu Ulumu Hatai
Mukaddimesi* (Turkish tr.), Istanbul, 1939.

scientists at both ends of the Old World came to be greater than it had ever been. Thus in the thirteenth century, we find that when a large astronomical observatory and a library of over 400000 volumes was established by the Arabs at Marāghah in Azerbaidjan, Chinese astronomers were there to collaborate. Arab medical work, too, contains evidence of Chinese collaboration, for the great Islamic physician of the early fourteenth century, Rashīd al-Dīn al-Hamdānī, wrote a book which not only contained much Chinese medicine but also a strong recommendation that the Chinese ideographic language should be used for science, since the meaning of its words was independent of pronunciation and therefore unambiguous (Fig. 21). There was transmission in the opposite direction, too, as evinced in the fascinating account of the visit by a Chinese scholar to the famous physician and chemist al-Rāzī. The Chinese, whose name we do not know, stayed with al-Rāzī for about a year, learned to speak and write Arabic, and about a month before returning to China, said he would like to copy out the sixteen books of the Graeco-Roman physician Galen, then the basis of all Arabic and Western medicine. This the scholar did, so quickly as to amaze al-Rāzī and his students, by using a form of shorthand that enabled him to write at a speed faster than they could dictate. But something must have happened to him on the way home for one can find no Galenic elements whatever in traditional Chinese medicine.

In spite of all the contacts that Islam made, those scholars who brought the science of the ancient Western world to Latin Christendom by way of translations from the Arabic sources concentrated solely on the works of Greek and Roman authors, neglecting all those Islamic works concerned with science in India and China. Of the sixteen important Arabic and Persian books that we know were written on these subjects, six still remain untranslated even today, and only one was translated before 1700. Western science therefore developed on the whole without the benefit of either Indian or Chinese contributions. Technical inventions, on the contrary, show a slow but massive infiltration from East to West for the first fourteen centuries of the Christian era, and include such epoch-making innovations as paper, printing, gunpowder and the magnetic compass, to mention only the classical discoveries noted by Francis Bacon in the seventeenth century.

General remarks

Of all the river-valley civilisations of antiquity, China was unique in its geographical isolation. In spite of this, and notwithstanding the difficulties in making contact, we can see that there was a virtually continuous diffusion of techniques to the West, if not of scientific ideas.

This transmission was of vital importance for the development of Europe. With civilisations less complex and less advanced than now, independent invention was less likely and transmission more important. A combination of gears to make a device like a hodometer for measuring distance travelled by a vehicle would have been a work of genius in the third century B.C. while it could be readily devised by any young mechanic today.

Yet although transmission was paramount, it is still frequently difficult to award priorities. To take an extreme example, the German astronomer Joseph von Fraunhofer invented in 1842 a special clock to drive a telescope so that it could follow the stars continuously in spite of the rotation of the Earth, and thus make observing more convenient. He did not know that even though they had no telescopes, the Chinese had done this eight centuries earlier with their own astronomical instruments. Was this development truly a re-invention? Again, there are other apparently independent discoveries that one feels convinced were really due to transmission even though we have no absolute proof. Suspension-bridges with wrought-iron chains are a case in point: first constructed in China in the sixth century A.D., they soon had successors in that part of the world, especially Tibet and other Himalayan countries, but they did not appear in Europe until the eighteenth century. Was this a case of independent invention or delayed diffusion? Here, as in so many other cases, dates of transmission are hard if not impossible to find, and we cannot be sure, though we do know that some of the European engineers knew of the Chinese bridges before any were built in Europe itself. Nevertheless it is clear that a host of technical devices – the wheelbarrow, the piston-bellows, the cross-bow; the technique of deep borehole drilling, the art and mystery of cast iron – were all known in China before, and often long before, they were known in the West. On the other hand, the Chinese also had to wait a very long time for some basic inventions to penetrate from the West, e.g. the screw and the crankshaft, to name only two; and there were some Chinese inventions which were known in the West but not adopted: paper money, the use of coal, and the adoption of water-tight compartments in ship-building for example. The whole situation of transmission of inventions and techniques has many facets, as Tables 6 and 7 show.

Lastly there is the question of what has been called 'stimulus diffusion', where an idea is transmitted without any details of the technique. The windmill is just such a case. An eighth-century Persian invention, it was always mounted horizontally, and it was in this form that it was introduced into China at the beginning of the Mongol period some five centuries later. From the first, however, the European windmill was

Table 6. *Transmission of mechanical and other techniques from China to the West*

	Approximate lag in centuries
(*a*) Square-pallet chain-pump	15
(*b*) Edge-runner mill	13
Edge-runner mill with application of water-power	9
(*c*) Metallurgical blowing-engines, water-power	11
(*d*) Rotary fan and rotary winnowing machine	14
(*e*) Piston-bellows	*c.* 14
(*f*) Draw-loom	4
(*g*) Silk-handling machinery (a form of flyer for laying thread evenly on reels appears in the eleventh century A.D., and water-power is applied to textile mills in the fourteenth)	3–13
(*h*) Wheelbarrow	9–10
(*i*) Sailing-carriage	11
(*j*) Wagon-mill	12
(*k*) Efficient harness for draught-animals:	
Breast-strap (postilion)	8
Collar	6
Variolation	1–7
(*l*) Cross-bow (as an individual arm)	13
(*m*) Kite	*c.* 12
(*n*) Helicopter top (spun by cord)	14
Zoetrope (moved by ascending hot-air current)	*c.* 10
(*o*) Deep borehole drilling	11
(*p*) Iron casting	10–12
(*q*) 'Cardan' suspension	8–9
Clockwork escapement	6
(*r*) Segmental arch bridge	7
(*s*) Iron-chain suspension-bridge	10–13
(*t*) Canal pound-lock	7–17
(*u*) Nautical construction principles	> 10
(*v*) Axial rudder	*c.* 10
(*w*) Gunpowder	4
Firearms	4
(*x*) Magnetic compass	11
Magnetic compass with needle	4
Magnetic compass used for navigation	2
(*y*) Paper	12
Printing (block)	6
Printing (moveable type)	4
Printing (metal moveable type)	1
(*z*) Porcelain	11–13

Table 7. *Transmission of mechanical techniques from the West to China*

	Approximate lag in centuries
(*a*) Screw	14
(*b*) Force-pump for liquids	18
(*c*) Crankshaft	3

vertical, as fourteenth-century illustrations show, and it seems that what was transmitted here was the idea of wind-driven vanes but nothing more. The concept alone came through and European mill-wrights adopted their own techniques, using right-angle gearing, to put it into practice. It was indeed as if someone, perhaps returning from the Crusades, had reported that the Saracens had harnessed the wind to grind their corn, and left it at that. The technicians had to go on from there and, in doing so, followed a different path. Again, the windmill is not an isolated case: there are plenty of others that could be cited.

It was in a variety of ways such as these, directly or indirectly, by travels of merchants and ambassadors, by capture of prisoners, or immigration of deserters, that there was cultural interchange between East and West, an interchange which seems, from more recent research, to have been greater than ever previously supposed.

7

Confucianism

With these background preliminaries complete, we can now turn to consider the part which Chinese philosophy played in relation to the development of scientific thought. From the start it must be emphasised that we shall find considerable differences between the Chinese outlook and our own; between Western traditional ways of looking at the natural world and the ways customary in China. What these differences were will become clear in this and later chapters, but the essence of it may be summed up by saying that whereas European philosophy tended to find reality in *substance*, Chinese philosophy tended to find it in *relation*. As we shall see, this was to exert a profound effect on Chinese science.

To begin with we shall examine briefly the *Ju Chia*, the Confucians, because of their dominance over all later Chinese thought, then move on to their great rivals, the *Tao Chia* (Taoists), whose speculations about Nature lie at the basis of all Chinese science. We shall also glance at the *Fa Chia* (Legalists) who advocated an almost fascist authoritarianism, yet held all men equal under the law, the *Mo Chia* (Mohists), chivalrous military pacifists with an interest in scientific method and experiments arising from military techniques, the *Ming Chia* who spent their time on logical paradoxes and definitions, and, lastly, the *Yin–Yang Chia* (School of the Naturalists) which developed a philosophy that gave the earliest Chinese scientific thinking its fundamental and characteristic theories. Other matters, like the rise of a sceptical outlook, the impact of Buddhism, and the appearance of Neo-Confucianism, will also get brief consideration. But first to Confucius himself.

Confucianism appeared in the sixth century B.C. and is named after its founder, about whom there are many traditions, not all acceptable historically. We know at least that his family name was Khung, his given name Chhiu, and his 'distinguishing name' Chung-Ni, but he is always referred to by his title of honour as Khung Fu Tzu (Master Khung),

of which Confucius is the Latinised form. Born in 522 B.C. in the state
of Lu (now Shantung), he traced his descent from the Imperial house
of Shang, and spent his life developing and propagating a philosophy
of just and harmonious social relationships. From about 495 B.C. he spent
a number of years in enforced exile from Lu, wandering from state to
state with a group of disciples and conversing with feudal princes, ever
hoping for a chance to put his ideas into practice. For the last three years
of his life he was back in Lu, writing and instructing his students. In
479 Confucius died, his life an apparent failure; yet as it turned out his
influence proved, in the end, to be so great that he has often been
called 'the uncrowned emperor' of China.

Confucianism was a doctrine of worldly social-mindedness. It strove
for as much social justice as was possible in a feudal-bureaucratic
society. This was to be achieved by a return to the ways of 'the ancient
Sage Kings' – a use of legendary historical authority that led Confucius
to term himself a transmitter rather than an innovator. In a chaotic feudal
society, torn apart by wars between states, Confucius sought order. In
a society in which human life was cheap, where there was little law and
order save what each man could enforce by personal strength, armed
followers, or intrigue, Confucius preached peace and respect for the
individual.

> 'When you go forth, behave to every one as if you were
> receiving a great guest; employ the people as if you were
> assisting at a great sacrifice, [do not] do to others what you
> would not wish done to yourself, and give no cause for
> resentment either at home or abroad.'

He advocated universal education and taught that diplomatic and administrative positions should go to those best qualified academically, not
socially; in this sense he was revolutionary. The true aim of government,
he taught, was the welfare and happiness of all the people, brought about
by no rigid adherence to arbitrary laws but by a subtle administration
of customs that were generally accepted as good and had the sanction
of natural law. In early Confucianism, then, there was no distinction
between ethics and politics; government was to be paternalistic.

From all this there followed one conclusion that could have been
important for science: the existence of intellectual democracy. If, as
Confucius believed, every man was potentially educable, then every
normal man was potentially as good a judge of truth as any other.
Moreover, Confucius counselled suspended judgement when there was
doubt; scribes, for instance, should follow the good old practice of
leaving a blank space when copying rather than faking a character of

which they were unsure. Yet Confucianism was not scientific in outlook: the universe had a moral order and the proper study of mankind was man, not a scientific analysis of Nature. Certainly Confucius taught a rationalist system that was opposed to any superstitious or even supernatural forms of religion, but it was an outlook that concentrated interest on social questions to the exclusion of all non-human phenomena. The rational element that could have encouraged the growth of a scientific outlook was not allowed to do so.

The greatest and most influential disciple of Confucius was Mêng Kho or Mencius, China's 'second sage', sometimes also known by his honorary title Mêng Tzu. Born in 374 B.C., more than a century after his master's death, Mencius spent most of his life advising the rulers of Liang and Chhi. There was little new in his teaching, although he emphasised the democratic conceptions of Confucianism, claiming that the goodwill of the people was essential in government: rites and usages were made for man, not *vice versa*, and became bad practice if they degenerated into empty conventions.

Doctrines of human nature

The most interesting aspect of Mencius from the point of view of scientific thought was his doctrine of human nature. 'All men', he said, 'have a mind which cannot bear to see the sufferings of others', a view that he used as a basis from which to argue that the feeling of commiseration is essential to man, and that, in consequence, man's nature had a general tendency to good. Yet not all Mencius' contemporaries agreed with him; some contended that individuals were morally neutral, others that men were born naturally evil. These irreconcilable views are interesting because they show us that in China as early as the fourth century B.C. there were present in germinal form those vast controversies that have lasted right through European as well as Chinese history; controversies that, until our own time, have had to be argued without the benefit of any clearly defined understanding of organic evolution. Nor were Mencius' views accepted by all later Confucians: in the next century, for instance, Hsün Chhing argued that human nature had a tendency to evil rather than to good, and that everything depended on education taken in its widest sense, although he did concede that every human being had an infinite capacity for development in the direction of good. By Han times, four hundred years after Mencius, the dual doctrine of Master Hsün (Hsün Tzu) was taken even further, as epitomised by the sceptical Confucian thinker Wang Chhung, who claimed that human beings are born with mixed endowments of goodness and badness, but he denied that human nature tends towards one or other quality.

Gradually, then, the Chinese Confucians changed from a dogmatic to a more scientific approach to the nature of man, and by the thirteenth century a new synthesis was put forward. In 1235 in his book *Shu Pho* (Rats and Jade), Tai Chih, quoting other ancient books, wrote:

> 'People talk about human nature – some say it is good, others that it is bad. Generally they prefer Mêng Tzu's view and reject Hsün Tzu's. After studying both books I realised that Mêng Tzu is talking about the heaven–nature and what he calls the goodness of human nature referred to its [innate] uprightness and greatness. He wished to encourage it. This is what the *Ta Hsüeh* (Great Learning) calls "[developing] sincerity".
>
> 'But Hsün Tzu is talking about the matter–nature, and what he called the badness of human nature referred to its [innate] wrongness and roughness. He wished to repair and control it. This is what the *Chung Yung* (Doctrine of the Mean) calls "forceful checking"...
>
> 'Thus Mêng Tzu's teaching is to strengthen what is already pure, so that defilement tends to disappear of itself. While Hsün Tzu's teaching is to remove defilement actively. Both are equally helpful to later students.'

With remarks such as these it seems to us now that it must have been only a matter of time before the Chinese reached an understanding that nothing but a knowledge of biological evolution would bring. But Chinese culture never of itself reached this stage.

The Ladder of Souls

In the West, during the fourth century B.C., Aristotle adopted the word *psyche* (soul) for the principle which differentiated living from non-living substance. His further investigations led him to the conclusion that there were different kinds of soul, and as we have already seen (p. 58), he believed plants possessed a vegetative soul, animals vegetative and sensitive souls, and man possessed three – vegetative, sensitive and rational. It was an ingenious and effective way of describing the way living things behave, and it dominated biology for centuries.

The Confucians had a similar system with its own inherent hierarchy. Thus a century after Aristotle we find Hsün Tzu saying:

> 'Water and fire have subtle *chhi* (spirits) but not *shêng* (life). Plants and trees have *shêng* but not *chih* (perception); birds and animals have *chih* but not *i* (a sense of justice). Man has *chhi*, *shêng* and *chih* in addition to *i*; therefore he is the noblest of earthly beings.'

The fact that Hsün Tzu lived a century later than Aristotle could be taken to mean that the Chinese system was derived from the West but, as already remarked (see p. 58), this is unlikely: it was, after all, a century and a half before the opening of the Old Silk Road. What is important, though, is the difference in approach of China compared with the Western world: whereas the West characterised rational thought (the rational soul) as man's distinguishing feature, the Confucians were concerned with his sense of justice – a social attribute.

The Chinese developed this 'ladder of souls' to a still greater elaboration. Between the sixth and fourteenth centuries A.D. Confucians and Neo-Confucians discussed it in detail – apart from universal matter-energy (*chhi*) and universal organising principles (*Li*), planets were given a vital force (*shêng chhi*), and animals and man a blood energy (*hsüeh chhi*) and a perception associated with it; finally *liang nêng* (good instincts) were added to the *chih* of animals and man. In Ming times the fourteenth-century biologist Wang Khuei gave a ladder that included inanimate as well as animate things. The heavens, he claimed, have *chhi* (subtle spirits), and so, too, do rain, frost, snow and dew. The Earth has *chhi* + *hsing* (形, form), while other substances with form may have *hsing* (性, natural endowments) as well: thus herbs, wood and some minerals possess *chhi* + *hsing* (形) + *hsing* (性), the intercourse between Heaven and Earth giving rise to this combination. It generates also one of *chhi* + *hsing* (形) + *hsing* (性) plus an additional quality *chhing* (sensitivity), these being possessed by all animals. Man combines all the qualities enumerated so far plus *i* (sense of justice). Wang Khuei also classified the qualities of watery secretions and the excretions of animals, their fur, scales, feathers and shells. It was a highly developed scheme which seems to be more than just a development of the ideas of Hsün Tzu. But it still saw man's unique quality as his sense of justice, not his rationality.

The humanism of Hsün Tzu

Having discussed the growth and development of Confucian ideas about human nature and the ladder of souls, we must return once more to the third century B.C. and Hsün Tzu, because he exemplifies something very important – the ambivalent relationship of Confucianism to science.

On the one hand Hsün Tzu preached an agnosticism that even went so far as to deny the existence of spirits, but on the other hand he objected to any scheme of scientific logic such as the Mohists and Logicians tried to formulate. He appreciated that technical processes were useful because they had practical application but denied the importance of theoretical investigations. Hsün Tzu's scepticism was evident enough; in the days of widespread belief in demons and ogres, he wrote:

'When a person walks in the dark, he sees a stone lying down and takes it to be a crouching tiger; he sees a clump of trees standing upright and takes them to be standing men. Darkness has perverted his clearsightedness...

'South of the mouth of the Hsia River there was a man called Chüan Shu-Liang. In disposition he was stupid and timorous. When the Moon was bright and he was out walking he bent down his head and saw his shadow, and thought it was a devil following him. He looked up and saw his hair and thought it was a standing ogre. He turned around and ran...

'Whoever says that there are demons and spirits, must have made that judgement when they were suddenly startled, or at a time when they were not sure, or confused. This is thinking that something exists when it does not.'

Hsün Tzu also attacked the prevalent belief in physiognomy – in telling fortunes from a person's appearance – and on the face of it his scepticism ought to have been encouraging to early science. That it was not was caused by the excessive humanism of his approach. Sufficiently influenced by the Taoists to use their concept of the Order of Nature, Hsün Tzu nevertheless exalted *li*, the essence of rites, good customs, and traditional observances, into a cosmic principle. The detailed pedestrian processes of scientific logic were not for him. That the Taoists might favour, but Hsün Tzu replied:

'You vainly seek the causes of things;
Why not appropriate and enjoy what they produce?'

To neglect man and speculate about Nature was, he believed, to misunderstand the whole universe. And so it was that he struck a blow at science by emphasising its social context too much and too soon.

Confucianism as a religion

It was during Han times that Confucianism became the official doctrine of the bureaucratic society, and it was Han Kao Tsu, the first Han emperor, who in 195 B.C. offered important sacrifices at the Khung family temple in honour of Confucius. Later, in A.D. 59, the Emperor Han Ming Ti ordered sacrifices to him in every school in the country, thus taking the worship of Confucius away from the Khung family and out into the state, transforming Confucius from a model for scholars into the patron saint of the scholar-officials. Confucianism became a cult, a religion, based on a kind of hero-worship and borrowing both from the cults of nature-deities and ancestor worship. But since the conception of a priesthood was inimical to Confucian thought, the guardians and celebrants of the new religion were the local scholars and high officials.

The state religion of China which grew up from the beginning of the Imperial Age was something rather different. It involved the position of the emperor as the high priest of all the people, in which capacity the annual sacrifices at the altars of Heaven and Earth, the Sun and Moon, the temples of Agriculture, and so on, devolved upon him. There had been precedents for this in the sacrifices at the altars of the soil and the grain which each of the feudal princes had made in his own realm, but the liturgies celebrated by the Emperor of All Under Heaven far outshone them in magnificence. The Confucians approved of all this, and doubtless participated as clerks and courtiers, but it was not Confucianism. Nor did it have anything much to do with the Chinese people as a whole, except in theory.

Confucianism has little connection with the history of science. A religion without theologians, it had no one to object to the intrusion of a scientific view on its preserves, but in accordance with the ideas of its founding fathers, it turned its face away from Nature and the investigation of Nature, to concentrate a millennial interest on human society, and human society alone.

8

Taoism

We shall now look at the world through the eyes of the Taoists, the 'irresponsible hermits', as their avowed opponents, the Confucians, called them. Even today their thought-system occupies a place in the background of the Chinese outlook at least as important as Confucianism, and some appreciation of it is vital for the understanding of Chinese science and technology.

The Taoist system was a unique mixture of philosophy and religion that also incorporated magic and a primitive or proto-science; it was the only system of mysticism the world has ever seen that was not profoundly anti-scientific. Its name, Taoism, was not derived after some founding father, as Confucianism was, but from the aim of its followers to seek a Tao or 'Way', the ancient pictograph for Tao being made up of a head together with a sign for going. Gradually, though, the word became a technical term, charged with philosophical and spiritual meaning, and as such untranslatable: 'Way' is but a shadow of its full meaning. One could call it the 'Order of Nature', or the immanent power within and behind the universe. But to avoid any misunderstanding it will be as well if we adopt the word 'Tao' just as it is: what it implies will become increasingly evident as we go along.

Taoism had two origins. First there were those philosophers of the Warring States period who followed a Tao of Nature rather than a Tao of Human Society, who did not want to be employed by the feudal princes, and who withdrew into the countryside or the wilderness to meditate and to study Nature. These were men who seem to have had a deep-seated feeling – though they were never able to express it fully – that human society could never be brought into the kind of order the Confucians hoped for without a far greater understanding of the natural world. The other origin of Taoism lay among the magicians and shamans who had entered Chinese culture from the north at a very early date, and played an important part in it as representatives of a kind of nature-worship and magic closely connected with popular beliefs. Shamanism was a religion which venerated the secondary divinities and spirits in the world of Nature, and believed that priests, by means of ritual trance, ecstasy, and imagined aerial voyages, could command these spirits, make contact with the unseen powers, cure sicknesses of body and mind in Man, and ensure good hunting and good harvests.

It may seem puzzling that two such different elements could combine, as they later did, to form a Taoist religion, but the problem can be resolved to a great extent when it is remembered that, in the earlier stages of civilisation, magic and science are indistinguishable. Only much later, when there is sufficient experimental evidence and enough scepticism, can a separation be made. Before such a time, the careful mixture and heating of substances for some alchemical experiment appeared to be little different from doing the same kind of thing in a witches' cauldron in order to cast spells. In the West this separation of magic and science was not achieved until the seventeenth century; in traditional China it was never fully achieved. But that is not important here; what is significant is that in seeking the Tao of Nature, the Taoist philosopher was moved to carry out experiments himself, and whatever their purpose, he believed in using his hands as well as his mind – alchemy was mainly a Taoist pursuit in China for this very reason. Thus the fact that the Taoist quest involved the philosopher in manual operations, highlights another radical difference between Confucianism and Taoism. No Confucian scholar would have sullied himself with manual work of any kind, whereas to the Taoist it was part of his search for the Tao. His acceptance of it at once carried him outside the charmed circles of feudal aristocratic philosophy, and of the bureaucratic ethos that followed it as well.

Unfortunately, Taoist thought has been much misunderstood if not ignored by Western translators and writers. Taoist religion has been neglected, Taoist magic written off wholesale as nothing more than superstition, and Taoist philosophy interpreted as purely religious mysticism and poetry. The scientific side of Taoism has been largely overlooked, and the political position of the Taoists misconceived. Certainly religious mysticism was present in early Taoist thought, and the most important Taoist thinkers should assuredly be numbered among the most brilliant writers and poets in history. And it is wrong to think of them merely withdrawing from the courts of the feudal princes just to avoid the battles between Confucian humanism and Legalist authoritarianism: on the contrary they launched bitter and violent attacks on the whole feudal system, a fact that has been passed over by most Chinese, as well as Western, expositors of Taoism. In brief, although Taoism was both religious and poetic, it was also at least as strongly magical, scientific, democratic and politically revolutionary.

The Taoist conception of the Tao

The Tao was the Way; not the way of life within human society, but the way the universe worked. It was the Order of Nature. This is shown clearly in the *Tao Tê Ching* (Canon of the Virtue (Power) of the Tao), written some time about 300 B.C. and perhaps the most profound and beautiful work in the Chinese language. It is traditionally ascribed to Lao Tzu, one of the most shadowy figures in Chinese history, but who seems probably to have lived during the fourth century B.C. There, speaking of any natural object, it is said:

> 'The Tao gave birth to it
> The virtue [of the Tao] reared it
> Things [within] endowed it with form,
> Influences [without] brought it to its perfection.
> Therefore of the ten thousand things [i.e. all that there is]
>   there is not one that does not worship the Tao and do
>   homage to its Virtue. Yet the worshipping of the Tao, and
>   the doing of homage to its Virtue, no mandate ever
>   decreed.
> Always this [adoration] was free and spontaneous.
> Therefore [as] the Tao bore them, the Virtue of the Tao reared
>   them, made them grow, fostered them, harboured them,
>   fermented them, nourished them and incubated them – [so
>   one must]
> Rear them, but not lay claim to them,
> Control them but never lean upon them,
> Be chief among them but not lord it over them;
> This is called the invisible Virtue.'

These notes that Lao Tzu strikes we shall hear again and again. The Tao, the Order of Nature that brought all things into being and governs their every action, does so not by force but by a kind of natural curvature of space and time. By yielding, and not imposing his preconceived ideas on Nature, Man will truly be able to observe and understand, and so to govern and control. Chuang Chou, a famous Taoist, in another important Taoist book, the *Chuang Tzu* – probably about the same date as the *Tao Tê Ching* – wrote:

> 'How [ceaselessly] Heaven revolves! How [constantly] Earth
> abides at rest! Do the Sun and the Moon contend about their

respective places? Is there someone presiding over and directing these things? Who binds and connects them together? Who causes and maintains them, without trouble or exertion? Or is there perhaps some secret mechanism, in consequence of which they cannot but be as they are?'

This is a kind of naturalist approach that emphasises the unity and spontaneity of the operations of Nature. In the *Chuang Tzu* there is an imaginary conversation between Lao Tzu and Confucius that shows quite clearly that to the Taoist, the biological world no less than the inorganic comes under the operation of the Tao.

Another element frequently to be met with is the physical and mental benefit to be obtained from following the Tao; an element that was later on to crystallise into the search for a kind of material immortality in which the body would be preserved and rarified so as to take its place among the *hsien* – immortal beings with youthful bodies who lived in the wild places of the Earth perpetually. This was a search that involved drugs and alchemical preparations, as well as yogistic breathing exercises, sexual techniques and gymnastics, which we shall discuss later.

In his study of Nature, the Taoist became aware, of course, of change, of action and reaction, and of still wider philosophical implications that their view of the natural world brought in its train. Thus in the *Chuang Tzu*:

'The Yin and the Yang reflected on each other, covered each other and reacted with each other. The four seasons gave place to one another, produced one another and brought one another to an end...Then were seen now safety, now danger, in mutual change; misery and happiness produced each other; slow processes and quick jostled each other, and the motions of collection [or condensation] and dispersion [or rarefaction, scattering] were established...Those who study the Tao [know that] they cannot follow these changes to the ultimate end, nor search out their first beginnings – this is the place at which discussion has to stop.'

This passage is of particular interest for two other reasons. First it claims that the ultimate beginning and end of things is the secret of the Tao, and that all men can do is to describe and study the natural world. This is, indeed, nothing less than a profession of faith in natural science, implying a rejection of metaphysics, of creation myths or speculation about the end of the world: an attitude that is echoed elsewhere in Taoist writings.

The second point of interest is that it shows the Chinese were aware of the physical processes of condensation and rarefaction – in other words of differences in density. Certainly the Greeks knew about this three centuries earlier, but transmission of such ideas was very unlikely before the first century B.C., so this third-century statement in the *Chuang Tzu* book gives us good reason to see it as independent Chinese thinking. Moreover, further study of Taoist books makes it evident that this was no isolated case; on a number of occasions they independently reached conclusions quite similar to those of the Greeks. They too entertained the idea that water is the basic element of the universe, as the *Kuan Tzu* of about 330 B.C., supposedly the views of the seventh-century philosopher Kuan Chung, makes clear:

'The Earth is the origin of all things, the root and garden of all life... water is the blood and breath of the Earth, flowing and communicating [within its body] as if in sinews and veins.'

The unity and spontaneity of Nature

If there was one idea the Taoists stressed more than any other, it was the unity of Nature, and the uncreated and eternal nature of the Tao. The 'sage embraces the Oneness [of the universe], making it his testing-instrument for everything under Heaven'. Again, in the *Kuan Tzu* book:

'Only the *chün-tzu* (gentleman) holding on to the idea of the One can bring about changes in things and affairs. If this holding on is not lost, he will be able to reign over the ten thousand things. The *chün-tzu* commands things and is not commanded by things, for he has gained the principle of the One.'

Certainly a quotation like this contains an element of religious mysticism, as do other passages of a similar kind, but this is only because we are still at an early stage in the development of Chinese thought, before the differentiation between religion and science. Yet in spite of the mysticism, it is clear from all we know of the Taoists that this is an affirmation of the unity of Nature, a unity that now, in the twentieth century, lies, we know, like universal gravitation, at the very foundation of post-Newtonian natural science. What is more, looking back at the Taoists, it is not stretching the imagination to see in many similar passages the Taoist political ideal of an undifferentiated form of society.

Taoist writings also show other aspects of the unity of the Tao to be found in Nature and the universality of their concept of science. In the *Chuang Tzu* book we find:

'The Master [perhaps Lao Tzu] said, "The Tao does not exhaust itself in what is greatest, nor is it ever absent from what is least; therefore it is to be found complete and diffused in all things. How wide is its universal comprehensiveness! How deep is its unfathomableness!"'

And more imaginatively:

'Tungkuo Shun-Tzu said to Chuang Tzu, "Where is this so-called Tao?"
Chuang Tzu answered, "Everywhere."
The other said, "Please specify an instance of it."
Chuang Tzu said, 'Well, it is here in these ants."
Tungkuo replied, "That must be its lowest manifestation, surely?"
Chuang Tzu said, "No, it is in these weeds."
The other asked, "What about a lower example?"
Chuang Tzu said, "It is in this earthenware tile."
"Surely brick and tile must be its lowest place?"
"No, it is here in this dung also."
To this Tungkuo gave no reply.'

Thus nothing lies outside the domain of scientific enquiry, no matter how repulsive, disagreeable or apparently trivial it may be. This is a really important principle, for the Taoists, who were orienting themselves in a direction that would ultimately lead to modern science, were to take an interest in all kinds of things – in seemingly worthless minerals, wild plants, and animal and human parts and products – all of course disdained by the Confucians. As the *Kuan Tzu* book put it: 'The sage is like Heaven, he covers everything impartially; he is like Earth, bearing everything up impartially.'

This impartiality, this refusal to give attention selectively but to treat all things alike, this disinclination to make ethical judgements on the world of Nature or of Man, was anathema to the Confucian. But it was the essence of natural science. Science must be ethically neutral, and it is to the credit of the Taoists that they saw this, even though they were born into what was essentially an ethical society. Man could not be the measure of all things, as the Confucians taught.

To the Taoist, Nature was not only unified and independent of human standards, it was also self-sufficient and uncreated. Their key phrase was *tzu-jan* – spontaneous, self-originating, natural. As Lao Tzu put it, 'The ways of men are conditioned by those of the Earth, the ways of Earth by those of Heaven, the ways of Heaven by those of the Tao, and the

Tao came into being by itself (*tzu-jan*)', an expression of the basic affirmation of scientific naturalism. This was a view underlined by Chuang Chou when he described the noises caused by the wind as it sweeps between the trees, rushing through spaces that act like 'nostrils, mouth or ears':

> 'Tzu-Yu said, "The notes of Earth then are simply those which come from its myriad aperatures, and the notes of man may be compared to those [which issue from tubes of] bamboo – allow me to ask about the notes of Heaven?"
>
> 'Tzu-Chhi replied, "When [the wind] blows, the sounds from the myriad apertures are each different, and its cessation makes them stop of themselves [*tzu i*]. Both these things arise from themselves – what other agency could there be exciting them?"'

This was a naturalist approach that was indeed unusual in times when phenomena like this were most frequently explained as the voices of demons, dryads and spiritual beings. Later on *tzu-jan* came to be universally adopted when speaking of natural phenomena: 'All things have their natural tendencies [*tzu-jan chih shih yeh*]' as the writer of the *Huai Nan Tzu* book put it in the second century B.C., indicating clearly that by then, at least, the Taoists had come to appreciate the principle of cause and effect, even if they did not express it in language so formal as that of the Greeks.

But the Taoists possessed more than a simple mechanistic approach to Nature. In the fourth century B.C. Chuang Chou, for example, had a sense of non-mechanical causation. In a striking passage on the natural processes in animals or in the human body given in the *Chuang Tzu* book, he says:

> 'It might seem as if there were a real Governor, but we find no trace of his being. . . But now the hundred parts of the human body, with its nine orifices and six viscera, all are complete in their places. Which should one prefer? Do you like them all equally? Or do you like some more than others? Are they all servants? Are these servants unable to control each other, but need another as ruler? Or do they become rulers and servants in turn? Is there any true ruler other than themselves?'

These words are striking when one considers what is now known about the complex interrelations of nervous stimuli and reaction to them, or the mutual influences of the glands of the endocrine system. Yet to the Taoists neither man nor the universe needed a conscious controller. They thought of Nature essentially as an organism but without the need for demigods of any kind to operate it.

Paradoxically enough, this attitude is exemplified in Taoist parables about automata that inventors were supposed to have constructed in human guise but which, when opened up, revealed nothing but mechanisms. On the face of it such devices would appear to propound a purely mechanistic view of nature, but a quotation will make clear why this is not so. Taken from the *Lieh Tzu* book attributed to Lieh Yü-Khou, it is the most striking of all stories of this kind.

> '"Who is that man accompanying you?" asked the king.
> "That, Sir," replied Yen Shih, "is my own handiwork. He
> can sing and he can act." The king stared at the figure in
> astonishment. It walked with rapid strides, moving its head up
> and down, so that anyone would have taken it for a live
> human being. The artificer touched its chin, and it began
> singing, perfectly in tune. He touched its hand and it began
> posturing, keeping perfect time... The king, looking on with
> his favourite concubine and other beauties, could hardly
> persuade himself that it was not real. As the performance was
> drawing to an end, the robot winked its eye and made
> advances to the ladies in attendance, whereupon the king
> became incensed and would have had Yen Shih executed on
> the spot had not the latter, in mortal fear, instantly taken the
> robot to pieces to let him see what it really was. And, indeed,
> it turned out to be only a construction of leather, wood, glue
> and lacquer, variously coloured white, black, red and blue.
> Examining it closely, the king found all the internal organs
> complete – liver, gall, heart, lungs, spleen, kidneys, stomach
> and intestines; and over these again, muscles, bones and limbs
> with their joints, skin, teeth and hair, all of them artificial.
> Not a part but was fashioned with the utmost nicety and skill;
> and when it was put together again, the figure presented the
> same appearance as when first brought in. The king tried the
> effect of taking away the heart, and found that the mouth
> could no longer speak; he took away the liver and the eyes
> could no longer see; he took away the kidneys and the legs
> lost their power of locomotion. The king was delighted.'

Written probably in the third century B.C., this passage bears every mark of being essentially a declaration of faith in naturalist explanations of all phenomena including the behaviour of man. And as if to underline this, the *Lieh Tzu* book has elsewhere another curious story, this time about the semi-legendary physician Pien Chhio, who is supposed to have performed a heart transplant on two men, an operation that left their outward appearance the same while exchanging their minds.

Stories like these must belong to a very ancient mechanistic-naturalistic tradition that seems to have been present in both China and Greece at about the same time. For, interestingly enough, there is a third story in this genre – the flying wooden kite or bird made by Mo Ti – that not only displays a mechanistic view of flight but which is closely paralleled by a story of the Greek philosopher Archytas of Tarentum. Archytas worked about 380 B.C., the very date of Mo Ti's death, so yet again it appears that in both civilisations some thinkers at least were developing the same kind of scientific approach to the world.

To the Taoists, the watchword became 'Find the causes', and all through the political upheavals that followed the change from feudalism to feudal bureaucratism during the time of the First Unification, they managed to pursue it. But life was not easy; the forces of ethical Confucianism became overwhelmingly strong, and pressurised Taoism either to amalgamate with them or to be driven underground. As a result, some collaboration did occur, as can be seen, for instance, in the *Lü Shih Chhun Chhiu* (Master Lü's Spring and Autumn Annals), the first part of which was completed in 239 B.C. Although there is much scientific argument in the book, it usually ends, uncharacteristically, with some application to human society:

> 'All phenomena have their causes. If one does not know these causes, although one may happen to be right [about the facts], it is as if one knew nothing, and in the end one will be bewildered . . . The fact that water leaves the mountains and runs to the sea is not due to any dislike of the mountains and love for the sea, but is the effect of height as such . . .
>
> 'All this is so, too, with regard to the endurance or fall of states, and to the goodness or badness of individuals. For everything there must be a reason. Therefore the sage does not inquire about endurance or decay, nor about goodness or badness, but about the reasons for them.'

However, in spite of the Confucian lip-service at the end, the denial here of teleology, of design in nature, is Taoist. And it is a view that is strongly emphasised elsewhere in the book; for instance, in a story about a banquet at which the complacent speech of one of the elders is interrupted by a twelve-year-old *enfant terrible* who suggests that man is only one among other animal species, and that fish and game had no more been created for the benefit of man than man for the benefit of tigers. There might be collaboration with the Confucians, but the essential Taoist outlook remained unaffected.

The approach to Nature: the psychology of scientific observation

When Taoism became a religion in the third century A.D., and Taoist temples came to be built, it is significant that they were always known as *kuan*: the usual terms *ssu* and *miao* applied to the temples of other religions were never used for those of the Taoists. Yet *kuan* means 'to look': it combines the sign for seeing with a graph which, in its most ancient form, was a bird. The original meaning, then, was probably to observe the flights of birds, perhaps with a view to making predictions from the omens obtained from them. But as time passed, the meaning changed; between 430 and 250 B.C. it had come to signify 'watch-tower', while it was also the regular word for observing natural events for divination. The use of *kuan* for the Taoist temple was a recognition of the essential Taoist approach to life, an approach that might have overtones of magic and divination, but one that was based on observation of the natural world.

This approach to the world of Nature is quite different from that needed for the management of society; it demands a receptiveness, a passivity, not a commanding activity. True observation also demands freedom from preconceived theories, in contrast to an attachment to a set of social convictions. The dichotomy between Taoism and Confucianism was fundamental, and epitomised by the Taoist symbolism of water and the feminine, a symbolism that has perplexed and misled many Western commentators. Yet writings like the *Tao Tê Ching* as early as the fourth century B.C. make this clear:

> 'The highest good is like that of water. The goodness of water
> is that it benefits the ten thousand creatures, yet itself does
> not wrangle, but is content with the places that all men
> disdain. It is this that makes water so near to the Tao.'

Water is yielding, it takes the shape of whatever container it fills, it seeps through invisible crevices, and its mirror-like surface reflects all Nature. Great rivers and seas gain their kingship over the lesser streams by being lower so that the streams flow into them: so, too, the sage, to be above the people, must speak as though he were lower than they are. The Taoist principle of leadership is leadership from within.

> 'He who knows the male, yet cleaves to what is female,
> Becomes like a ravine, receiving all things under heaven.

[Thence] eternal virtue never leaks away.
This is returning to the state of Infancy.'

A quotation from the *Tao Tê Ching*, but only one of many typical examples of the Taoist appreciation of the female principle, and the receptive passivity of science.

This attitude was fundamental and led the Taoists intuitively to the roots of science and of democracy. Whereas the Confucian–Legalist ethic was masculine, hard, managing, domineering, even aggressive, the Taoists emphasised a totally different approach – feminine, tolerant, yielding, permissive, mystical and receptive. They opposed a feudal society since theirs was a poetical expression of a collectivist society, a society of the type that had existed in the early village communities before the social differentiations of the Bronze Age. If it were not unthinkable from the Chinese point of view, that the Yin and the Yang, the female and male principles, could ever be separated, it would be tempting to describe Taoist thought as a Yin system and Confucian as a Yang one. But since the indivisibility of the Yin–Yang principle prevents this – a point recognised by every Chinese philosopher, Taoist or otherwise – a new word was introduced to cover it, the term *jang*. The dictionary definition of *jang* is to yield up, to cede, to give up the better place, to invite. It is an expression of the ancient idea of 'potlatch', where the prestige of a leader depends on the amount of food or other commodities he can distribute to the community as a whole at periodical feasts. In China the magical virtue, the social prestige, the 'face', derived from ceding and yielding, became a dominant element: and not only in Taoist circles: indeed, as Needham has remarked, it is still evident to anyone who has lived in China and has experienced the difficulties of passing through a doorway with a group of people, or seen scholars, or for that matter high party bureaucrats, positively struggling for the least honourable places at a dinner-party. Yet though not exclusive to Taoism, it reached its greatest expression in Taoist philosophy, leading Taoists often enough to refuse state office when this came their way.

With the attitude of the Taoist observation of Nature in mind, we can now turn to the question of their motive for doing so. Once again, the answer is to be found in Taoist writings, and we have to go no further than the *Lieh Tzu* book to find a useful parable. A man of Chhi ' was so afraid that the universe would collapse and fall to pieces leaving his body without lodgement, that he could neither sleep nor eat', but the remedy Lieh Yü-Khou advocated was not to pity his distress, but to enlighten him about the facts of Nature. Only in this way would he gain peace of mind. And this parable is no joke, as many have taken it to be, but a story to underline the need for rational explanation in the face of the more terrifying manifestations of the natural world. Earthquakes, eruptions, floods, epidemics, can all strike terror into the mind of man, but as soon as he has begun to examine and classify different types of catastrophe, as soon as he has started to take a detached scientific look at the events, he immediately feels stronger and more confident. Speculating on their cause, discussing their nature, making an informed guess about their future occurrence, and giving them technical names, can provide a peace of mind to be got in no other way. Reflective thinkers like the Taoists, who were not content with the Confucian concentration on the affairs of human society, stood in need of assurance: the observation of Nature provided it.

This confidence and inner calm obtained by contemplating the workings of the natural world was known to the Chinese as *ching hsin*, and to the Greek followers of the atomic ideas of Democritus and Epicurus as *ataraxia*. The parallel is close and unmistakable. Thus in the *Chuang Tzu* book:

> 'The ancients who regulated the Tao nourished their
> knowledge by their calmness, and all through life refrained
> from employing that knowledge in action [contrary to
> Nature]; moreover they may also be said to have nourished
> their calmness by their knowledge.'

Compare the *De Rerum Natura* (On the Nature of Things) of Lucretius (about 65 B.C.), where the Epicurean atomic-scientific philosophy is set out in verse:

> 'These terrors, then, this darkness of the mind
> Not sunrise with its flaring spokes of light
> Nor glittering of morning can disperse,
> But only Nature's aspect, and her Law.'

Chuang Chou wrote movingly on this topic and the quotation from his book is only one of many. He frequently refers to 'Riding on the Normality of the Universe' and to the 'Infinity of Nature'; he describes the sense of liberation which can be gained by those who can abstract themselves from the trivial quarrels of human society and unify themselves with the great world of Nature. And Chuang Chou was echoed again and again by other Taoists. Indeed the *ching hsin*, the inner peace that brought the sage impassability, ran right through Taoism. In the eighth century A.D. we still find it; in the *Kuan Yin Tzu* book:

'Minds occupied with fortune and misfortune may be invaded and controlled by devils. Minds occupied with love affairs may be attacked by lustful ghosts...Minds concentrated on drugs and tempting food [literally 'baits'] may be attacked by the ghosts of material things...Only the sage can control the spirits and not be controlled by the spirits. Only he can make use of all things, grasp their mechanisms, connect all things, disperse all things, defend all things. For every day the sage faces the facts of Nature, and his mind is untroubled.'

Taoist contemplation brings peace. What of Taoist action? In the *Chuang Tzu* it is said that those who nourished knowledge by their calmness refrained from employing their knowledge 'in action contrary to Nature'. At least this is how Joseph Needham believes the words *wu wei* should be translated, for he rejects the customary translation implying inaction. In other words the Taoist must refrain from going against the grain, from trying to make things perform functions for which they are unsuitable, from attempting to exert force in human affairs when the man of insight can see that it would be doomed to failure. And his view is not without support. For instance, in the *Huai Nan Tzu* (The Book of (The Prince of) Huai Nan) of 120 B.C., Liu An says:

'Some may maintain that the person who acts in the spirit of *wu wei* is one who is serene and does not speak, or one who meditates and does not move; he will not come when called nor be driven by force. And this demeanour, it is assumed, is the appearance of one who has obtained the Tao. Such an interpretation of *wu wei* I cannot admit. I never heard such an explanation from any sage...What is meant, therefore, in my view, by *wu wei* is that no personal prejudice [or private will] interferes with the universal Tao, and that no desires and obsessions lead the true courses of techniques astray. Reason must guide action in order that power may be exercised according to the intrinsic properties and natural trends of things.'

Again, about A.D. 300, in a commentary on the *Chuang Tzu* book, Kuo Hsiang was equally specific: 'Non-action does not mean doing nothing and keeping silent. Let everything be allowed to do what it naturally does, so that its nature will be satisfied.' Thus, if the passages in Taoist writings about *wu wei* are translated in Needham's sense, they all fit in well with this attitude, and the gem-like brevity of the fourth-century B.C. *Tai Tê Ching* becomes clear: 'Let there be no action [contrary to Nature] and there is nothing that will not be well regulated.'

As we shall see later, this interpretation of *wu wei* also echoes the deepest roots of the archaic nature of primitive peasant life: plants grow best without interference from Man, men thrive best without interference from the state. It is not inactivity but harmony with Nature that is the aim. The Han Confucians, however, did not appreciate this view which, after all, was contrary to everything they taught. Consequently, they emphasised the quality of effortlessness in *wu wei* which, under the influence of Buddhist meditation techniques in later Han times, did come to mean the avoidance of every kind of activity. As a result, gradual misunderstanding of *wu wei* grew, and in the end led to abuses that brought Taoism into discredit.

Taoist empiricism

The ability to practise *wu wei* implied learning from Nature by observation. This, in its turn, brought the Taoist to a scientific approach; by an almost imperceptible transition, it led him to experimentation, and was of capital importance for the whole development of science and technology in China. A remarkable statement on this empirical outlook is to be found as early as the third century B.C. in the *Lü Shih Chhun Chhiu*:

> 'To know that one does not know – that is high wisdom. The
> fault of those who make mistakes is that they think they know
> when they do not know. In many cases phenomena seem to
> be of one sort [alike] when they are really of quite different
> sorts... Lacquer is liquid, water is also liquid, but when you
> mix the two things together, you get a solid. Thus if you
> moisten lacquer it will become dry. Copper is soft, tin is soft,
> but if you mix both metals together they become hard. If you
> heat them they will again become liquid. Thus if you wet one
> thing it becomes dry and solid; if you heat a [hard] thing it
> becomes liquid. Thus one may see that you cannot deduce the
> properties of a thing merely by knowing the properties of the
> classes [of its components]...
>
> 'In the state of Lu there was a man called Kungsun Cho
> who said he could raise the dead. When they asked him how,
> he replied, "I can heal hemiplegia [apoplexy]. If I gave a
> double dose of the same drug, I could therefore raise the
> dead." But among other things there are some which can have
> small-scale effects, but not large-scale ones, and other things
> which can perform the half but not the whole.'

Practical experience was obviously important in this context, and with this in mind the text goes on to show that the craftsman is sometimes more knowledgeable than the supposedly learned man.

'Kaoyang Ying was having a house built. His mason said, "It won't do to use wood that is too green; when it is plastered it will warp. If you use fresh wood, the house may look all right for a short time, but it will be certain to fall down before long." Kaoyang Ying replied, "[On the contrary] according to your own statement, the house cannot fall down. The drier the wood the harder it will be, the drier the lime the lighter it will be. If you put something which is always getting harder with something which is always getting softer they could not possibly hurt one another." The mason did not see how to answer this, so he accepted the order and built the house. When it was done it looked well, but very soon it fell to pieces. Kaoyang Ying liked such small sophistries, but had no understanding of the great principles [of Nature].'

The mason knew better than his sophist employer because practical experience had taught him the true nature of the materials he used. However well the rationalist might argue, Nature would win in the end, confounding him and justifying the empirical attitude of the Taoist. And this instinctive stance long continued: for we still find it in the eighth century A.D., for instance, when Han Kan, the greatest painter of horses during the whole Thang dynasty, is said as a youth to have preferred to spend his time in the imperial stables rather than accept the emperor's offer to be instructed by the most notable painters of the day. The importance of empirical knowledge, whether of artisan or sage, was a Taoist theme that was to echo and re-echo throughout the centuries in Chinese thought.

Change, transformation and relativity

With so concentrated an interest in Nature, the Taoists were bound to become intimately concerned with the questions of change and transformation. And they were not alone in this; the Yin–Yang Chia (Naturalists) and the Ming Chia (Logicians) also pondered on the subject. Various types of change were classified: changes caused by previous actions, changes that were part of a cyclic process whereby everything returns, in due course, to the beginning again; changes that were gradual and changes that were sudden. In particular, the Taoists also recognised inward and outward changes. They saw that even the Taoist sage himself would experience change, adapting himself to the world of Nature as experience dictated, yet without forsaking his fundamental outlook. Needless to say, though, they concentrated in changes in the natural world, and were especially intrigued by cyclic change, not only of the seasons, but also as evident in all kinds of cosmic and biological events. In this, once again, they travelled along similar lines to the Greek atomists.

When it came to the question of life and death, the Taoist had a tendency to incorporate an appreciation of, and a resignation to, the change. The *Chuang Tzu* book comments:

> 'Since death and life thus attend upon each other, why should I account [either of] them an evil?...[Life] is accounted beautiful because it is spirit-like and wonderful. [Death] is accounted hateful because it is foetid and putrid. But the foetid and putrid, returning, is transformed again into the spirit-like and wonderful; and then the reverse change occurs once more.'

Here and elsewhere there is an appreciation of change as part of the understanding of Nature and a resignation to it that lay at the root of the Taoist calmness of mind. For the *Chuang Tzu* book was not alone amongst Taoist writings in seeing the end of one phase as the beginning of something else.

Here a feeling of contradiction might emerge between this calm Taoist philosophical resignation to inevitable change, decay and death, on the one hand, and that equally strong Taoist religio-scientific quest for the means of material immortality on the other. But really there was no contradiction, because longevity and immortality were to be gained by particular techniques, i.e. ways of working that went along with Nature, not in opposition to her. The only problem was to find them out. In Taoist thought there was also the important proviso that the sage recognises that truth may be distributed among many opinions: views that are propounded will never be totally correct, or totally in error, and can only be judged at all from the 'axis' of the Tao. The harmonising of conflicting opinions can only be accomplished in the light of the invisible operations of Heaven, i.e. the facts of history and Nature. This is amusingly illustrated in the *Chuang Tzu* book with the famous parable of the monkeys:

> 'To wear out one's spirit and intelligence in order to unify things without knowing that they are already in agreement – this is called "Three in the Morning". Why? A keeper of monkeys once said with regard to their rations of nuts that

each monkey was to have three in the morning and four at night. But at this the monkeys were very angry. Then the keeper said they might have four in the morning and three at night, and with this arrangement they were all well pleased. His two proposals were substantially the same, but one made the creatures angry and the other glad. Thus the sages harmonise the affirmations "it is" and "it is not", and rest in the natural equalisations of Heaven. This is called "Following two courses at once".'

There is thus a true dialectical quality in Chuang Chou who, in his views of change as eternal, and reality as a process, has much in common with the nineteenth-century European philosopher Hegel and his Marxist successors. It is a quality that is found throughout Taoist thought, and interesting from the scientific point of view since it led the Taoists to deny that animal species were fixed, coming close accordingly to a theory of evolution. Indeed, with this in mind we can now discern the meaning of a chapter in the *Chuang Tzu* book that has been the despair of translators; a chapter that begins:

'All species contain [certain] germs [*chi*, minute seeds]. These germs, when in water, become *chüeh* [minute organisms]. In a place bordering on water and land they become [lichens or algae, like what we call the] "clothes of frogs and oysters". On the bank they become *ling-hsi* [probably a kind of plant]'

and continues, moving from plants to insect larvae, from larvae to butterflies and crabs, then to birds, horses, and finally Man. Here, clearly, Chuang Chou is describing Taoist observations on insect metamorphosis and their extension of these to cover a whole continuous chain of linked animal forms.

The idea of change and transformation of animal species – of evolution – also led the Chinese to the suggestion that different aptitudes had arisen in response to different environments and thus nearly to a concept of natural selection. Indeed, the Taoists were particularly impressed by the great differences between forms and functions: what was good for one species could be bad for another. They did, in fact, even appreciate that the use or uselessness of a species might be an advantage in its survival: in many passages the value of being useless is commented on – trees, for instance, attain great size and longevity only when they are of no use to anyone and so avoid being cut down. Certainly there may be here, and in similar passages, an echo of the Taoist desire to withdraw from active and social life, but nevertheless the arguments do represent some understanding of the concept of survival of those best fitted to their environment.

Another significant scientific aspect of Taoist thought was the appreciation of a certain degree of relativity. Man-oriented judgements were seen to be absurd when applied to the non-human world: for instance the time-scales of animals and plants were bound to be very different from those of man. Thus in the *Chuang Tzu* book:

> 'Small knowledge is not to be compared with great, nor a short life to a long one. The morning mushroom does not know what happens between the beginning and end of a month, the *hui-ku* [mole-cricket] knows nothing of the alternation of spring and autumn. These are instances of short spans of life. But south of Chhu State there is the *ming-ling* [a species of tree] whose spring is five hundred years and whose autumn is equally long... And among men Phêng Tsu [the Chinese Methuselah] was especially renowned for his length of life – if all men were to match him, would they not be miserable?'

There is here, and in other passages that could be quoted, a refusal to make distinctions of quality between great and small. This refusal may have had political overtones, but that does not lessen its scientific value, for in dealing with large and small, the Taoists were well aware of the relativity of appearance, and of optical illusions. The *Lü Shih Chhun Chhiu* says:

> 'If a man climbs a mountain, the oxen below look like sheep and the sheep like hedgehogs. Yet their real shape is very different. It is a question of the observer's standpoint.'

Noteworthy, too, is the fact that this Taoist appreciation of change and relativity was coupled with their rejection of man as the centre of all things, an orientation that lay at the very root of Confucianism.

The Taoist interest in change was not, however, solely scientific: some of their ideas about it were clearly magical, as we should expect in view of the closeness between science and magic in ancient times. Yet in spite of magical elements, and the fact that the Taoists themselves never developed a systematic natural philosophy, they were quite adept at developing the practical consequences of their observations. As the *Chuang Tzu* book says:

> 'It was separation [that led to] completion; and from completion [ensued] dissolution. But all things, without regard to their completion and dissolution, return again into

the Unity [of Nature]. Only the far-reaching in thought can know how to comprehend them in this Unity. This being so let us give up devotion to our own preconceived notions, and follow the "common" and "ordinary" views, which are grounded on the *use* of things. The study of that *use* leads to comprehension, and that secures success. That success gained, we are near [to the object of our search] and there we [have to] stop. When we stop, and yet do not know how it is so, we have what is called the Tao.'

The spirit of technology without a theoretical, or at least a fully theoretical, background is evident here, and with this it is no wonder that great strides were made. The Taoists clearly did not fall into the trap of believing that full theoretical understanding was necessary before any technological advance could be achieved; they realised that, in the light of experience, the technologist will often manage to do the right thing in a given situation, even if sometimes he does it for the wrong theoretical reasons.

The attitude of the Taoists to knowledge and society

We must now face squarely the question of the political position of the Taoists, an aspect that has been greatly misunderstood in the West, no less so than the proto-scientific and observational tendencies described so far. Essentially, the Taoists 'walked outside society', with an attitude to knowledge which was quite different from that of the Confucians and the Legalists. Confucian social knowledge, 'the difference between princes and grooms' as Chuang Chou put it, seemed nonsense to the Taoist, to whom real knowledge was knowledge of the Tao and of Nature, not man-made doctrines of social classes. To obtain true knowledge the Taoist 'emptied his mind' of all 'false' Confucian-style knowledge, neglecting distorting memories, prejudices and preconceived ideas. He encouraged empiricism, and respected the technology of craftsmen, an attitude that was to have profound practical effects since it encouraged the great inventors of ancient China. All this was, of course, the very antithesis of the Confucian outlook, rooted in totally different political and ethical connections.

The close relationship in the Taoist mind between political position, a practical approach to problems, even their attitude to knowledge and their mysticism, were not unique. If we turn our glance for a moment to the West at the time of the rise of modern science during the Renaissance, we shall see something of a parallel. Modern science arose here not only because of dissatisfaction with ancient ideas in the light

of experience, but also because of the way ancient ideas were allied with
the Establishment. Christian theology had allied itself closely with the
scientific theories of the ancients, and particularly with the teachings
of Aristotle, and it was with this powerful body of established doctrine
that nascent modern science had to contend. The new scientific
fraternity wanted to overthrow the old out-dated ideas in favour of
experimental results and a new view of age-old questions, yet to do this
meant attacking the whole edifice of established thought. However, the
new experimental empiricists were not on their own; they found unlikely
allies in the Christian mystics. Unlikely, that is, in today's terms, but
not so when it is remembered that this was all occurring at a time in
the West when Christianity dominated men's minds, the basic cultural
background of both the Establishment and its opponents. Christianity
was bound to appear on both sides of the struggle. Christian mysticism
allied itself with the new scientific movement because it, too, embodied
a certain degree of empiricism: the rational theologians did not, and so
found themselves naturally on the side of the ancient order.

   This division in the West, like the Taoist–Confucian clash in the East,
also had political overtones. In northern Europe, where the scientific
revolution was most successful, many of the new scientific practitioners
were on the Protestant–Puritan side with all its political implications for
the new social order. The outstanding mystical naturalist Paracelsus
(1493–1541) shows us in extreme form much of this early revolutionary
spirit. Standard-bearer of alchemy applied to medicine, proponent,
against all opposition, of mineral drugs, first observer of the occupational
diseases of miners, he was an experimentalist and a theoretician. He was
also politically equalitarian. An intense individualist, he saw the
salvation of society in a kind of collectivism, and unwittingly echoed
Chuang Chou when he said it was not God's will that there should be
lords and commoners; all men were brothers. Paracelsus had much in
common with the Taoists. Indeed it can be shown that his alchemical
medicine derived ultimately from the elixir concept of China mediated
through Arabic and Byzantine culture.

   This Western parallel with the East is not unique: similar situations
occurred elsewhere. In Islam mystical theology was closely allied in the
tenth century with the developments of science in Persia and Mesopotamia: a semi-secret equalitarian society, the Brethren of Sincerity,
was close to Taoism both in its politics and its scientific interests. Again
like the Taoists, the Brethren moved over to religious mysticism when
they found it impossible to put their socialist aims into practice.

   The Taoist aim for society was a kind of agrarian collectivism,
without feudalism and without merchants; they advocated what was

virtually a return to a simpler way of life. This was a reaction against the Confucian and Legalist control of society, an authority that treated common men as instruments, considering some men noble and others of little worth. While the Taoists disapproved strongly of Confucian–Legalist attitudes, they did consider it proper to 'rob' Nature for the good of the community, to use the natural world for the well-being of man; what they did consider wrong was to 'rob man' to accumulate wealth for private ends. The Taoists, then, wanted an equalitarian society and looked to the simple communal spirit of the past as a guide, a tribalism that was probably originally matriarchal – one of the reasons, perhaps, why the feminine symbol was so dear to them. For a time they may even have believed that a return to primitive society was possible, and sought for rulers who would put their principles into practice, but like other reformers since, they had no success.

Since their aim was a return to an earlier and 'purer' state of society, the Taoists were not revolutionaries in the usual sense of the word. But in this, and in other matters, their political attitudes have been much misunderstood, mainly due to a lack of appreciation of certain technical terms that they used, most particularly the Chinese words *phu* (the uncarved block) and *hun-tun* (chaos). An examination and re-translation of passages in the *Chuang Tzu* and the *Huai Nan Tzu* books, as well as older writings, make it clear that whatever the later meanings of *phu*, in the second and third centuries B.C. it contained a political element; then *phu* referred to the social solidarity of each man with his neighbour. The word *hun-tun*, or the words *hun* and *tun* separately, meant 'undifferentiated' and 'homogeneous', that is classless in the Taoist social context. And these interpretations, mainly unappreciated in the West, are underlined and confirmed by those peculiarly Chinese figures of Taoism, the 'Legendary Rebels'. These were characters whom the earliest legendary kings were supposed to have fought and conquered; and the names with their political overtones make it evident that some were personifications of political ideas. Some had names with a distinctly similar ring to the political terms just mentioned. Thus the legendary rebel Huan-Tou was identified with the monster Hun-Tun banished by the 'Yellow Emperor' Huang Ti. But the name Huan-Tou itself means, literally, 'peaceable bellows', and the legendary monster was supposed to have been the inventor of metallurgy and metal weapons; another monster, Thao-Wu, had a name which means 'untrimmed stake, post, beam or log', and another Kung Kung, chief of the artisans, a name that means 'communal labour'. Such names clearly show primitive connections with the working people, and it is likely that the legendary rebels were the leaders of a pre-feudal collectivist society, their banishment symbolising the change to feudalism. Such an interpretation also fits in with other legendary beings – the Three *Miao* and the Nine *Li* – who seem to represent metal-working confraternities that existed before the days of feudal society.

If, then, this interpretation of the Taoist political outlook is correct – and it seems to make sense of much that was previously obscure in Taoist writings – then we should expect to find close connections between Taoists and the mass of the people, the manual workers. Such connections do exist, more particularly with the fourth-century Taoist philosophers Hsü Hsing and Chhen Hsiang, of whom Mêng Tzu, a century later, gives us glimpses in a description of co-operative agricultural units. And as if to confirm the truth of this concern of the philosopher with manual work, Taoist writings not infrequently have much to say about manual skills, especially where some knack or 'flair' is involved. In the early days of any technology, flair is crucial, and the Taoist interpretation of it was that it could not be taught or transferred: it could only be attained by minute concentration on the Tao running through natural objects of all kinds. This interest of the philosopher in manual skill was, of course, utterly foreign to Confucianism, but it fitted in well with the equalitarian beliefs of the Taoists.

Although they deeply respected manual skill, adopted an empirical outlook, and stood close to the ingenuity of mechanics and inventors, the Taoists nevertheless also expressed at times a paradoxical distrust of technological innovation. At first sight this seems very curious, in direct opposition to their naturalist philosophy and their known connections with science and technology; and as with their attitude to knowledge, it has led many commentators off on a false scent. Yet a careful examination of their expressions of distrust makes it clear that what the Taoists were objecting to was the misuse of technology, not technology itself; to its use as a means of enslavement of men by the feudal lords. This is all epitomised in a story in the *Chuang Tzu* book where Tzu-Kung talks to a farmer who is drawing water from a well with a bucket. He explains that there is a very simple labour-saving device, the counter-weighted swape (or *shadūf*), for doing this, but the farmer laughs and replies:

'I have heard from my master that those who have cunning devices use cunning in their affairs, and that those who use cunning in their affairs have cunning hearts. Such cunning means the loss of pure simplicity. Such a loss leads to restlessness of the spirit, and with such men the Tao will not dwell.'

'I knew all about the swape, but I would be ashamed to use it.'

The reason for the Taoist view is not hard to find. The power of feudalism rested partly on control of certain crafts like bronze-making and irrigation engineering, while the differentiation of classes had gone hand in hand with technical invention; in consequence, the inventions themselves seemed of dubious value, not because bad in themselves but because they were so easily used for the wrong purposes. For example, any kind of tool or machine could be used for purposes of torture by the feudal lords, and therefore it was natural that Taoist defenders of the people should look askance at it.

The Taoists can be said, then, to have hankered after a certain primitivism, looking back to a Golden Age. In the West parallel views prevailed from time to time – the repudiation of civilised life by the Stoics and the Cynics of Greece, the Christian doctrine of primitive bliss before the Fall of Man, the eighteenth-century admiration of the 'Noble Savage'. Yet although the West had groups with such ideas, they never quite matched the Taoists, who were more organised than the Stoics and Cynics, and whose combination of political anti-feudalism coupled with the beginnings of a scientific movement had no Western equivalent.

Shamans, 'wu', and 'fang-shih'

In describing Taoism and the Taoist attitudes we must be careful not to neglect its connections with the primitive religion and sorcery of the North Asian peoples, mentioned at the beginning of this chapter, and particularly with the shaman, that priestly spirit-possessed healer and magician, who appears in China as a *wu*. The name *wu* is significant since the word is connected with dancing, the character found on oracle-bones actually depicting a dancing wonder-working shaman holding plumes or feathers in his hands, or in hers, for a woman could be a *wu*. At once this has a special significance, for Taoism was not only a mystical philosophy acknowledging the magical practices of the shamans, but also connected with the use of the feminine symbol generated by primitive matriarchal society. Another word, *fang-shih*, is also important: it has been translated as 'a gentleman possessing magical remedies' and is clearly connected with the healing aspects of shamanism. Considering the intimate connection between shamanism, exorcism, and early medicine, the use of *fang-shih* in a Taoist relationship is understandable enough, but there is an even closer connection, for the Taoists saw a definite link between *wu*, pharmaceutics, and their study of alchemy.

The *wu* were not always on good terms with authority, so that the political as well as the magical aspects of the Taoist outlook found echoes here. Indeed, by Sung times the *wu* were sternly persecuted by governors and prefects, and down to the end of the Chhin dynasty provisions against sorcerers and wizards remained in the Penal Code. But since the *wu* had at times carried out such practices as human sacrifice, one can have some sympathy with the Confucian opposition, at least on humanitarian grounds. In any case this suppression drove the *wu* aspect of Taoism underground, and led in due course to the formation of secret societies among the people, societies that were to play a large part in Chinese life in later centuries.

The aims of the individual in Taoism

As briefly mentioned at the beginning of this chapter, Taoism was captivated from the earliest times by the idea that it was possible to achieve material immortality, i.e. persistence not in some other world, but on the Earth. Since men possessed souls the Chinese believed that it was not possible to have continued existence without some form of bodily component, the string on which they were threaded, as it were. The Taoists, in fact, believed in the existence of a band of holy immortals (*shêng hsien*). Fascinated by youth they felt sure that processes to prevent growing old could be found. By mediaeval times the factors causing ageing had become 'personified' into the 'Three Worms' and the 'Three Cadavers', and immortality techniques were concerned with getting rid of these. Only then could a man become a *chenjen*, a 'True Man', living on for ever with a youthful if etherealised body. To achieve this desirable state special funeral rites had to be performed over the body of the dead adept, who would have devoted a lifetime of preparation involving a number of special practices – respiratory, heliotherapeutic, gymnastic and sexual.

The respiratory techniques and breathing exercises go back to high antiquity in China. Their aim was to enable the adept to return to the manner of respiration in the womb, and since the Taoists knew nothing of the gases present in the circulation of the mother or her foetus, they interpreted this by trying to make the breathing as quiet as could be and to practise holding the breath for as long as possible. Holding the breath led to all kinds of effects – buzzing in the ears, vertigo, sweating – and these were all thought to be good training for the immortal state. The second immortalising technique, heliotherapy or the use of sunbathing (not recognised as valuable in Europe until modern times), was practised by exposing the body to the sun while at the same time holding in one's hand a special character (the sun within a border) written in red on green paper. However, this was done only by the men; women adepts had to expose themselves to the moon, holding a piece of yellow paper with the moon within a border drawn in black – a technique that could hardly have enriched their vitamin D content.

The third technique, the gymnastic exercises, were called *tao yin* (extending and contracting the body), and may have been derived from the rain-bringing dances of the shaman, though the Indian yogistic postures exerted a great influence too. Later names were *kung-fu* and *nei kung*, implying 'work' or 'inwardly-directed work'. All were derived from the age-old idea that the pores of the body were liable to become obstructed, thus causing stagnation of body fluids and disease. Massage techniques were also used.

The fourth practice, the use of sexual methods, met with great antagonism from both Confucians and Buddhists, but is nevertheless of considerable interest today. In view of the general acceptance of Yin–Yang theories, it was natural to think of human sexual relations as having intimate connections with the mechanism of the whole universe, and Taoists regarded them as an important aid to material immortality. Indeed, some of the ancient texts on the subject give the names of sexological experts who were noted for their longevity. Techniques practised in private were called 'the method of nourishing life by means of the Yin and Yang'. Their basic aim was to conserve as much as possible the seminal essence (*ching*) and to make use of the two great forces in individuals as indispensable nourishment for each other. No sharp line of distinction can be drawn between Taoist practice and customary behaviour, although in keeping with Taoist outlook, the importance of women in the scheme of things was greatly emphasised. It is probable that some of the techniques derived from situations in patrician families where there were many concubines, and this is not surprising since the problem of organising a healthy sex-life in a polygamous household must have been a very real one.

The Taoist techniques really combined two opposites – sexual stimulation to increase the *ching*, and methods to prevent its loss. Continence was considered against the rhythm of nature, and celibacy (advocated later so strongly by the Buddhists) as leading to neuroses, so that to avoid loss of the *ching* a frequently used method was *coitus reservatus*, making love with a succession of partners but ejaculating rarely. Today such a series of unfulfilled intromissions might be considered psychologically harmful, but to the Taoist it was different because the aim was different. The technique was not practised for a negative purpose – to avoid conception – but for a positive reason – to ensure the mutual nourishment of the two forces, and especially to strengthen the male

Yang. Another method was to exert a pressure on the urethra between scrotum and anus at the moment of ejaculation, thus diverting the seminal fluid into the bladder. Thence, the Taoists thought, the *ching* could be sent upwards to 'nourish the brain' (*huan ching pu nao*). They did not know that in this *coitus thesauratus* the fluid was later lost by ordinary excretion. There are Indian parallels for this also.

Much emphasis was laid on a succession of partners, with many (and conflicting) directions for their choice, but since there was also an elaborate system of prohibitions that depended on the seasons, the weather, the phases of the moon, the astrological situation, and so on, suitably propitious occasions for the Taoist adepts tended not to occur very often. However, the most astonishing aspect of this side of Taoist physiological-religious practice was that it comprised public ceremonies in addition to ordinary conjugal life and private exercises for the adepts. The ceremonies originated during the second century A.D. and were common by 400: they consisted of a liturgical dance ending either in a union of the two chief celebrants in the presence of the congregation, or successive unions of the members of the assembly in chambers along the sides of the temple courtyard.

Taoism as a religion

In 1943 Joseph Needham made an excursion with a number of eminent scientists from Kunming, the capital of Yunnan province, to the western hills to visit three temples there, two Buddhist and one Taoist. In view of the scientific interests of ancient Taoist thought, all were especially keen to visit the 'Chamber of the Three Pure Ones', a rock-hewn shrine built half-way up an almost perpendicular cliff. However, Needham soon discovered that not one of the party had the slightest idea who the 'Three Pure Ones' were, and he remarks that this typifies the lack of study given to one of the most interesting phenomena in the whole of comparative religion. For our part, we cannot leave Taoism without glancing at this subject, for we need at least some explanation of the disappearance of those germs of scientific thought that were so prominent in ancient and early mediaeval Taoism, and some idea of how it all became transformed into an organised liturgical theistic religion.

To begin with, there is no doubt that religious Taoism was a reaction against the collective religion of ancient Chinese feudal society with its altars to the gods of the soil and grain. For as the state enlarged and state religion grew, it became increasingly impossible for the majority of the people to participate in the rites, with the result that Taoism became China's indigenous individualistic religion of salvation. It began during Han times and owed much to the Chang family of the first

Fig. 22. Taoist temple Wu-liang Kuan at Chhien-shan, south-east of Anshan, near Shanyang, Liaoning. Original photo.

century A.D. Tradition has it that the family was descended from Chang Liang who, in the first century B.C., had helped the adventurer Liu Pang to gain power over the Chhin and their Legalist advisers whom the Taoists loathed. At all events, the family were very powerful and their development of Taoism into a religion would have been taken as an example to be followed. Thus when, a century later, another member of the Chang family, Chang Tao-Ling, a Taoist and alchemist, developed the nascent religion further, he acquired so large a following that he was able to set up a semi-independent state on the borders of Szechuan and Shensi. Chang Tao-Ling was believed to possess magical powers, and soon after his death, in 165, Taoism had increased so much in prestige that official imperial sacrifices were offered for the first time to Lao Tzu.

The activities and doctrines of Chang Tao-Ling may have received some stimulus from abroad, but whatever the causes, the second century A.D. saw the rise of a definite Taoist Church, which flourished during many subsequent centuries. Then, in 423, at the court of the Northern Wei dynasty, the Taoist Khou Chhien-Chih assumed the title *Thien Shih* (Heavenly Teacher), and founded what has sometimes been called the Taoist papacy, which descended in an unbroken line right down to the present century. However, the Taoist Church itself underwent many upheavals as the years passed, most notably in the form of controversies with the Buddhists, and in due course these struggles exhausted both sides. Thus their influence waned and intellectual prestige passed in Sung times (eleventh and twelfth centuries A.D.) to the Neo-Confucians. Then, when foreign dynasties like the Mongols and Manchus came to power, there followed a great mistrust of Taoism in official circles, not least because of its subversive political power in anti-foreign agitation, and its claims to divinatory powers that could so easily be used to launch a change of dynasty. Under such official disapproval the Taoist Church continued to decline even further.

During the times when Taoism was really flourishing as a religion, divine revelations were received by many of its leaders. These started during the third century A.D. and continued in subsequent times, and so in the fifth century, there arose the doctrine of a Trinity – the Three Pure Ones. These were known as the Precious Heavenly Lord and First Original Venerable Heavenly One, controlling time past; the Precious Spiritual Lord and Great Jade-Imperial Heavenly Venerable One, controlling time present; and the Precious Divine Lord, the Pure Dawn Heavenly One, controlling time to come. This doctrine of three persons may possibly have owed its existence to Christian influence, but was rather early for this unless some Gnostic ideas came through. More probably it was a consequence of Taoist ideas of cosmogony which as early as the fourth century B.C. had envisaged that 'The Tao produced the one, the one produced the two, the two produced the three, and the three produced the ten thousand things [i.e. everything].'

When one looks at the whole picture, one is inescapably led to the conclusion that Taoism developed into a Taoist Church as an indigenous alternative to the ecclesiastical organisation of the Buddhists. Yet this did not drive Taoist philosophy, with its political collectivism, its mysticism, and its scientific attitudes, altogether into oblivion. In fact, between them, Taoism and Confucianism still form the background of the Chinese mind, and will continue to do so for a long time to come. Indeed, the roots of Taoism are vigorous even now, epitomised, perhaps, in the Taoist temple set in beautiful gardens at Kunming where, after ascending through the lower halls with their various images, one comes at last to an empty hall where there is nothing but a large tablet inscribed *Wan Wu chih Mu*, the Mother of All Things.

9

The Mohists and Logicians

Here we have two schools of Chinese thought which tried hard to work out a basic scientific logic, the Mohists doing so with strong political bias, the Logicians only to a lesser degree. Mohism, however, was completely overwhelmed by events at the end of the Warring States period; so effectively, in fact, that we do not even know the exact dates of the birth or death of its founder Mo Ti. All we can say for certain is that his life fell wholly within the period 479 B.C. to 381 B.C., and that he died not long before the birth of Mencius, who wrote against him. A native of the state of Lu, he may have been a minister in Sung, and certainly seems to have kept a school for those who wished to become officials to the feudal princes. His great doctrines, which have made him one of China's noblest figures, were those of universal love, and the condemnation of offensive, but not defensive, war.

The Mohists have been said to represent the 'chivalrous' element in Chinese feudalism, for although they condemned war, their pacifism had its limits; indeed, they trained themselves in the military arts so they could hurry to the aid of a weak state attacked by a strong one. And it was their concern with the techniques of fortification and defence that seems to have led them to their interest in the basic methods of science. Their studies in mechanics and optics are among the earliest records of Chinese science in existence, and if the Taoists had a special interest in biological change, the Mohists can be said to have been attracted mainly to mechanics and physics. This has not always been appreciated and scholars have concentrated on the ethical passages in the *Mo Tzu* book, the compendium of Mohism, at the expense of the scientific propositions.

Mo Ti's religious empiricism

The Taoists, as we have just seen, preached against feudalism. They believed social evolution had taken a wrong turn and they wanted to return to a more primitive state of social solidarity. The Mohists, on the other hand, condemned primitive society, where, they said, every man had followed his own dictates and there had been continual dissension. There was 'disorder in the human world' like that to be found 'among animals and beasts', all due 'to the want of a ruler'. On the other hand they did not take the Confucian attitude: rather they taught that once the world became one community, there would be harmony and love of one's fellow men, and concern with their well-being would then prevail. Kindness and compassion would become general, coupled with a basic collectivism. All this was lost, the Mohists believed, when the empire became a 'family inheritance', when men concentrated on their own family units at the expense of the community, when, in fact, a thoroughgoing Confucian ethic took precedence over any other attitude.

The way out of the situation was to return to the outlook of the sage-kings of old who were selfless: only then could a 'Great Togetherness' (*Ta Thung*) result, and this, the Mohists taught, meant the practice of *chien ai* or universal love. Their aim was a practical one – to make feudalism work better – and in this sense they possessed a Confucian outlook, but it was tempered very strongly by a religious ethic which condemned the exploitation of the weak. The Will of Heaven 'abominates the large State which attacks small States, the large house which molests small houses, the strong who plunder the weak, the clever who deceive the stupid, and the honoured who disdain the humble'.

The Mohists were extremely conscious of an unseen world; they believed in the existence of ghosts and spirits, whom they thought of as watchers of the morality of the living, and strangely enough, they were led to this view by their scientific empiricism.

> 'Mo Tzu said: "The way to find out whether anything exists or not is to depend upon the testimony of the eyes and ears of the multitude. If some have heard it or some have seen it then we have to say it exists. If no one has heard it and no one has seen it, we have to say it does not exist... If from antiquity to the present, and since the beginning of Man, there are men who have seen the bodies of ghosts and spirits, and have heard their voices, how can we say they do not exist?"'

This attitude of Mo Tzu was by no means unscientific; the appeal to the community of observers is part of the structure of natural science.

He arrived at the wrong conclusion because he underestimated the rôle of the critical intellect, as was pointed out five centuries later in a discussion on the Mohists by Wang Chhung, which appeared in the *Lun Hêng* (Discourses Weighed in the Balance) of A.D. 83:

> 'The fact is that truth and falsehood do not depend [only] on the ear and eye, but require the exercise of the intellect. The Mohists, in making judgments, did not use their minds to get back to the origins of things, but indiscriminately believed what they heard and saw. Consequently, although their proofs were clear, they failed to reach the truth.'

Scientific thought in the Mohist canon

Mo Ti is to be highly praised for his doctrine of universal love preached as early as the fourth century B.C., and paralleling the best teachings of the monotheistic religions of the West, but this has little direct connection with the history of science. Yet when we come to examine the Canons and Expositions in the *Mo Tzu* (Book of Master Mo), we realise how far the later Mohists went in their efforts to establish a thought-system on which experimental science could be based. Some selected examples may make this clear (C = Canon, CS = Exposition):

> 'C An attribute [literally a 'side'] may be [added on to or] taken away from [something] without involving increase or reduction...
> CS Both are the same one thing and no change has occurred.'

This refers to subjective judgements like a 'beautiful' flower, which remains the same flower, whether thought beautiful or not.

> 'C Fire is hot...
> CS Fire: when one says fire is hot, this is not [only] on account of the heat of the fire; it is [because] I make the assimilation [or correlation] [of the visual sensation of] light [and the tactile sensation of heat].'

The work of the mind in sorting and ordering sensations and perceptions was much discussed by the Mohists. Perception has for its object the world which the sense organs, the 'five roads', apprehend; their data are then subject to reflection, and it is through this that conceptual or interpretative knowledge (*chih* 怒) is attained. It is interesting that this character was apparently invented by the Mohists as a technical term, and has long since disappeared from dictionaries.

We turn now to 'models' or 'methods' of Nature:

'C The mutual sameness of things of one *fa* (model or
method) extends to all things in that class. Thus squares are
the same, one to another. . .
CS All square things have the same *fa*, though [themselves]
different, some being of wood, some of stone. This does not
prevent their squareness mutually corresponding. They are all
of the same kind, being all squares. Things are all like this.'

and next to causation:

'C A cause is that with the obtaining of which something
becomes [comes into existence].
CS Causes: a minor cause is one with which something may
not necessarily be so, but without which it will never be so.
For example, a point in a line. A major cause is one with
which something will of necessity be so [and without which it
will never be so]. As in the case of the act of seeing which
results in sight.'

The minor cause here is what we should call a necessary condition rather
than a cause, but there is no doubt that in a passage like this we find
ourselves in the very engine-room of scientific thinking. So too with the
question of what constitutes knowledge:

'C Knowing comprises hearing about something, making an
inference from it or an exposition of it, experiencing it
personally, harmonising names with actualities, and then
action. . .
CS Receiving something transmitted is hearsay knowledge.
[Classifying] unhindered by position in space [because the
things concerned may be far apart] is inference or exposition.
What is observed by one's own body is personal experience.
What designate are names, what are designated are actualities;
when names and actualities are yoked together like a plough-team,
that is [the required] harmony. So also will mated to
movement is action.'

There is absence here of any prejudice against action (*wei*), so characteristic of the Taoists.

'C When one hears that what is not known is like what is
known, then both are known. . .
CS What is outside is known. Then someone says, "The
colour inside the room is like this colour [outside]." Thus
what is not known is like what is known. . . Names serve, by

what is understood, to make certain what was not previously known. They do not use the unknown to conjecture at what is understood. It is like using a metre rule to measure an unknown length.

Then there is an important passage on 'knowledge and practice':

> 'C If one has a general idea which one does not as yet understand [what to do about it?] . . .
> CS A last, a hammer and an awl, are all things used for making shoes. The ornamentation may be put on before the shoe is hammered, or afterwards. The process takes its course; the exact order of the operations may be a matter of chance [the precedence may be equivalent].'

Only by practice can essentials and non-essentials be distinguished. Rational argument was recommended:

> 'C To hold that all speech is perverse, is perverseness . . .
> CS To hold that all speech is perverse is not permissible. If the speech of the man [who urges this doctrine] is permissible, then speech is not perverse. But if his speech is permissible, it is not necessarily correct.'

This quotation is more than a recommendation, however; it is an attack on the Taoist mistrust of reasoned argument. But not all Taoist ideas were contrary to Mohist teaching; the Mohists agreed, for instance, on the peace, the liberation from fear, that can be obtained by a study of Nature:

> 'C Calmness of mind is [the acquirement of] knowledge without preferences or attractions and without prejudices or repulsions.
> CS Calmness of mind: tranquillity [in the acceptance of] the thus-ness of things.'

These quotations show that, in spite of some similarities, we are in a different world from the Taoists. There is nothing of the Taoist poetry and vision, and there is less interest in life phenomena as such, although the Mohists recognised change, and in particular biological change. Yet even though we have to see their work through the dark glasses of corrupted texts and ingenious emendations, the broad sweep of their understanding, the way they sketched out what amounts to a complete theory of scientific method, is what matters; and this shines through even the distorted records. They discussed sensation and perception, causality and classification, agreement and difference, and the relations of parts and wholes. They recognised the social element in fixing a terminology, and distinguished between first-hand and second-hand evidence. Yet unfortunately they never proposed a general theory of natural phenomena more satisfactory than the Five-Element theory (to be discussed in the next chapter): they criticised the doctrine in detail, but that seems to have been as far as they could go.

The Mohists tried to define various forms of scientific reasoning, and even though the texts are somewhat uncertain, it seems reasonably established that they reached the two important and basic principles of deduction and induction. The first is to be seen, for instance, in a passage about the use of mental 'models':

> '"Model-thinking" consists in following the methods [of Nature].
> What is followed in "model-thinking" are the methods.
> Therefore if the methods are truly followed by the "model-thinking" [literally: hit it in the middle], the reasoning will be correct.
> But if the methods are not truly followed by the "model-thinking", the reasoning will be wrong.'

In other words, if one correctly recognises the causes (by formulating a 'model') then one arrives at the correct answer, and since in Nature these causes will be far fewer in number than the effects being considered, determining the causes can only be by a process of deduction.

Again, for induction, for reasoning from the particular to the general, the following passage is valuable:

> 'Extension is considering that that which one has not yet received [i.e. a new phenomenon] is identical [from the point of view of classification] with those which one has already received, and admitting it.'

Here, indeed, we have a generalisation being formed from a limited number of instances.

With their conceptual models, their deduction and induction, the Mohists, like the Greeks, reached the very threshold of the theory of science. Indeed, it is tempting to think that if Mohist logic and Taoist naturalist insight could only have been combined, the Chinese might have crossed that threshold. The tragedy of Chinese science is that this never happened.

The philosophy of Kungsun Lung

Never very clearly differentiated from the Taoists and Mohists, the Logicians or 'School of Names', the *Ming Chia*, were listed separately in the first century A.D. by the astronomer and historian Ssuma Than and the historian Pan Ku. Its two greatest names were Hui Shih, who lived during the fourth century B.C., and Kungsun Lung, whose life fell mostly in the first half of the next century. Both were contemporaries of the Taoist Chuang Chou who, after the death of Hui Shih, complained that there was no longer anyone with whom he could talk. Hui Shih and Kungsun Lung were both consultants to feudal princes in the manner of Warring States scholars, both had students whom they tried to interest in their logical exercises, and both, it seems, had little success. All their writings are lost except for a partially preserved book, the *Kungsun Lung Tzu*, and the paradoxes recorded in the *Chuang Tzu* and elsewhere.

The *Kungsun Lung Tzu* has been said to reach the highest point of ancient Chinese philosophical writing. It is in dialogue form, like many of Plato's writings, and the part that has survived is concerned with what are now known as 'universals' (e.g. 'white', 'horse', 'hard', etc.) as distinct from particular things. A typical and simple discussion of universals (*chih*) runs as follows:

> 'There are no things [in the world] that are without *chih*, but these *chih* are without *chih* [i.e. they cannot be analysed further or split up into other *chih*].
> 'If the world had no *chih*, things could not be called things [because they would have no manifested attributes]...
> 'There are no [materially existing] *chih*, [and yet it has been stated above that] there are no things that are without *chih*. That there are no *chih* [materially existing] in the world, and that things cannot be called *chih*, does not mean that there are no *chih*. It is not that there are no *chih*, because there are no things that have not *chih*...
> 'That there are no *chih* existing in the world [in time and space], arises from the fact that all things have their own names, but these are not themselves *chih* [because they are individual names, not universals].'

Again, omitting the dialogue form of the original:

> 'A white horse is not a horse...The word "horse" denotes a shape, "white" denotes a colour. What denotes colour does

not denote shape. Therefore I say that a white horse is not a horse [as such] . . . When a horse [as such] is required, yellow and black ones may all be brought forward, but when one requires a white horse, they cannot. . . Therefore yellow and black horses are things of the same kind, and can respond to the call for a horse, but not to the call for a white horse. Hence it results that a white horse is not a horse [as such, or horseness].'

The apparent absurdity – that a white horse is not a horse – was stated to attract prospective thinkers, and was one of a number of the Logicians' paradoxes. Yet from the point of view of natural science, the greatest interest focuses on the discussion of change, the central problem in the investigation of Nature.

'Q Does two contain one?
A Two (-ness) does not contain one.
Q Does two contain right?
A Two (-ness) has no right.
Q Does two contain left?
A Two (-ness) has no left.
Q Can right be called two?
A No.
Q Can left be called two?
A No.
Q Can left and right together be called two?
A They can'

The universal of two is simply twoness and nothing else. But 'right' added to 'left' is two in number, and they can therefore be called two.

'Q Is it permissible to say that a change is not a change?
A It is.
Q Can "right" associating itself [with something] be called change?
A It can.
Q What is it that changes?
A It is "right".'

In other words this is the universal of righthandedness manifesting itself in double things, then disappearing again.

'Q If "right" has changed, how can you still call it 'right'?
    And if it has not changed, how can you speak of a change?
A "Two" would have no right if there were no left. Two

contains 'left-and-right'. A ram added to an ox is not a
horse. An ox added to a ram is not a fowl.'

Here Kungsun Lung is clearly aiming to show that the universal is
changeless, while the particular is ever changing.

The paradoxes of Hui Shih

The writings of the Logicians always contain a desire to shock,
as, for instance, when they write that all quadrupeds have five legs each,
although their serious aim is to draw attention to the unchanging
universal 'quadruped-leg-as-such'. However, in his writings Hui Shih
was no mere sophist – he had a great interest in science as well as logic
– and his paradoxes were always designed to bring out certain logical
points. They were, indeed, similar to the famous paradoxes of the Greek
philosopher Zeno of Elea. Zeno lived about a century before Hui Shih,
but once again the difficulties of transmission make it likely that the
Greek and Chinese philosophers arrived at the paradox method
independently.

Some of Hui Shih's paradoxes were connected with the all-pervading
nature of change and relativity. Thus:

'The Heavens are as low as the Earth; mountains are on the
same level as marshes'

brings out the point that there must be a frontier at which the heavens
touch the Earth, and the fact that, from the point of view of the
universe, the irregularities of the Earth's surface are minimal. Again:

'The South has at the same time a limit and no limit'

which has been variously interpreted either that there are regions beyond
the bounds of known geography, or that it indicates that the Earth was
known to be spherical, as perhaps does another paradox:

'I know the centre of the world, it is north of the state of Yen
[the most northerly Chinese state] and south of the state of
Yüeh [the most southerly state].'

Besides these paradoxes about the relativity of space, some were
concerned with another relative aspect, the relativity of time.

'The Sun at noon is the Sun declining, the creature born is
the creature dying'

applies both to astronomy and biology. Astronomically the brief moment
of noon seems illusory, and the Sun is always declining from some place

on the Earth's surface; biologically it is a reference to ageing which, according to modern science, goes on at its greatest rate the younger an organism is. Did Hui Shih hit the mark better than he knew? Then again,

'Going to the state of Yüeh today, one arrives there yesterday'

sounds like a phrase out of a textbook on Einstein's relativity, and surely shows a recognition of different time-scales at different places. Of course, these are not the only explanations – it could be, as an early twentieth-century Confucian scholar claimed, that these paradoxes were designed to show that all spatial distinctions are unreal and illusory – but on reflection it seems more likely that their aim was similar to that of Zeno, seeking to prove an underlying continuity in Nature as against a discontinuity.

Relativity of space and time, and of change, did not exhaust the subjects with which Hui Shih was concerned. He also had paradoxes in infinity related to a kind of atomism, on classification and universals, on the rôle of the mind, on potentiality and actuality, and on natural wonders that seemed paradoxical. As an example of the first:

'The greatest has nothing beyond itself, and is called the
Great Unit; the smallest has nothing within itself, and is
called the Small Unit',

the 'Great Unit' being, it would seem, the universe of space and time just mentioned, the 'Small Unit' the essence of an atom since an atom was something that could be divided up no further. Atomism was, however, a concept that soon came under attack by the late Mohists and Logicians, just as it had done with Zeno. The paradox

'A dog can [be?, become?, be considered as?] a sheep'

is a typical example of one on classification and universals, meaning, presumably, that both dogs and sheep are quadrupeds.

'Fire is not hot'
'Eyes do not see'

brings us face to face with what and how we perceive: fire is not hot of itself, although this is how the mind interprets it; eyes do not see by themselves, but are merely sense-organs serving the mind. Again a different type of paradox, one on potentiality:

'An egg has feathers'

emphasises the potential hatching of a chicken. And a paradox like:

'Mountains issue from mouths'

is typical of those concerned with natural wonders, here probably referring to the eruptions of volcanoes. Still other paradoxes deal with mathematical and mechanical questions, and with animal generation.

Logic, formal or dialectical?

Paradoxes of the type favoured by the Logicians are sometimes to be found in books which are generally considered Taoist. For instance, in the *Lieh Tzu* book a discussion is recorded between a philosopher Hsia Ko – who may or may not have been a real person – and the Emperor Thang, of the Shang dynasty. The discussion concerns antinomies, or contradictions between what appear to be equally logical conclusions, reminiscent of the eighteenth-century European philosopher Immanuel Kant.

> 'Thang of the Shang asked Hsia Ko saying, "In the beginning, were there already individual things?" Hsia Ko replied, "If there were no things then, how could there be any now? If later generations should pretend that there were no things in our time, would they be right?" Thang said, "Have things then no before and no after?" To which Hsia Ko answered, "The ends and the origins of things have no precise limits. Origins might be considered ends, and ends origins. Who can draw an accurate distinction between these cycles? What lies beyond all things, and before all events, we cannot know."
>
> 'So Thang said, "What about Space? Are there limits to upwards and downwards, and to the eight directions?" Hsia Ko said he did not know, but on being pressed, answered, "If there is emptiness, then it has no bounds. If there are things, then they have bounds. How can we know? But beyond infinity there must exist non-infinity, and within the unlimited again that which is not unlimited. [It is this consideration] – that infinity must be succeeded by non-infinity, and the unlimited by the not-unlimited – that enables me to apprehend the infinity and unlimited extent of space, but does not allow me to conceive of its being finite and limited.'

The two questions in this extract correspond to the first and second antinomies of Kant, but the emphasis on infinity is distinctly Taoist, although the manner of treating it is akin to the methods of the

Logicians. But for our purpose the important thing is that the work of the later Mohists and Logicians is of central importance in the study of the development of scientific thought in China.

The Mohists and Logicians attempted to lay foundations on which the world of natural science could be built. In doing so they show an unmistakable tendency towards dialectical rather than formal logic, expressing it in paradox and antinomy; a tendency that is typical of Chinese thought, always concerned with relationships rather than the problems raised by substance. Where Western minds asked 'What, essentially, is it?', Chinese minds asked 'How is it related in its beginnings, functions and endings with everything else, and how ought we to react to it?'

In this matter of dialectical and formal logic, it is worth noting that in the scientific awakening in Europe in the seventeenth century, there was strong attack on formal logic – the syllogistic reasoning codified by Aristotle, and the only system known during European mediaeval times. As Thomas Sprat, the historian of the new scientific movement in England, put it in the late 1660s:

> 'This very way of Disputing itself, and inferring of one thing from another alone, is not at all proper for the spreading of Knowledge... In brief, Disputing is a very good Instrument to sharpen Men's Wits, and to make them versatile and wary Defenders of those Principles which they already know: but it can never much augment the solid Substance of Science itself.'

Later, scientific men were to echo and re-echo this view, pointing out that such formal logic had retarded independent thinking; put in another way, formal logic was an inadequate tool when it came to handling the greatest fact of Nature, change, as indeed the Taoists had recognised. And this is why examples of dialectical or dynamic logic in ancient China are of such interest when we consider the development of science there.

Formal logic, it has been said, may have been a necessary stage in the development of science in Europe, but no one can be certain of this. Nor shall we ever know whether, had the environmental conditions of Chinese society been favourable for the full development of natural science, the Mohists or some other school might not have formulated such a logic. One thing is sure, the philosophers of the Warring States period used in their writings all the forms of reasoning catalogued by the Greeks, without naming or describing them in an abstract way. And there is another possible reason for this: some mathematical logicians today maintain that the principles of syllogistic logic were more perfectly incorporated in the Chinese monosyllabic ideographic language than in any Indo-European alphabetical one – in this case the Chinese would never have felt the need for a codification of reasoning forms.

We do not know for certain the causes of the decay and disappearance of the ideas of the Mohists and Logicians during the upheavals of the first unification of the empire. Probably social conditions polarised thought into two moulds, the Confucian and the Taoist. The specific social aims of the Confucians would preclude any close attention to logical problems – as the Confucian Hsün Tzu put it, 'the superior man does not discuss them; he stops at the limit of profitable discourse' – but nevertheless the Mohist ideal of universal love was absorbed into Confucianism during the Han and after, modifying the principle of graded affection enunciated by Mencius. The Mohist interest in science and technology that remained passed the Confucians by, but not the Taoists, into whose tradition it became incorporated.

If the disappearance of a separate Mohist school is accounted for in this way, what of the Logicians? It is sometimes said that their work was entirely unknown to the mediaeval Chinese, but this seems an exaggeration. The existence of the 'Name Principle' (Ming Li) school of the Chin period suggests that their kind of discussion was continuing during the third and fourth centuries A.D., and in the eighth century it was claimed, in a memorial to the recently deceased Taoist philosopher Chang Chih-Ho, that he had written a book 'Mystical Theses on Hardness, Whiteness and Horseness'. If this was so, then the subjects first discussed by Mo Ti were still a subject for Taoist scholars twelve centuries later. When one considers the enormous gaps known to exist in those ancient Chinese writings that have survived, and compares the Taoists, the Mohists and the Logicians with their Greek equivalents, one is left with the impression that there is little to choose between ancient European and ancient Chinese philosophy as far as the foundations of scientific thought are concerned.

10

The fundamental ideas of Chinese science

We now approach a field of vital importance for the history of scientific thought in China: the fundamental ideas and theories that were worked out from the earliest times by the Chinese naturalists. The subject conveniently divides into three sections: first, the theory of the Five Elements (*wu hsing*); secondly, that of the Two Fundamental Forces (*Yin* and *Yang*); thirdly, the scientific, or to be more precise, the proto-scientific use of that elaborate symbolic structure, the Book of Changes (*I Ching*). In the light of modern research, our discussion will differ considerably from the Chinese traditions taken over rather uncritically by the early Western sinologists, and it will be as well if we preface it by a glance at the origin and development of some of the Chinese words most important for scientific thought.

Origins of some of the most important Chinese scientific words
Before any science can develop there must be a suitable stock of words, and now, due to the discovery of the Anyang oracle-bones already described (Chapter 4), as well as the characters inscribed on Shang and Chou bronze vessels, there is a copious graphic vocabulary for study. Not all the characters have been identified, but even so there are enough for a selection to be made of those ideographs that throw light on the origins of Chinese scientific terms. It is true that the ancient words probably had little influence on the thinking of the exponents of the proto-sciences in Chhin and Han times, but for us these early ideographs are of interest in themselves because they help us to understand the ancient Chinese approach to science. A selection of some of the important words is to be found in the accompanying table (Table 8), where the English word is followed by the modern pronunciation of its Chinese equivalent, then the Chinese character, next its ancient form, and finally a brief explanation of its archaic significance.
A study of the table shows that the fundamental terms necessary for

Table 8. *Ideographic etymologies of some of the words important in scientific thinking*

No.	Word	Modern Chinese		Ancient oracle-bone, bronze, or seal form	K no. (see note, p. 141)	Remarks
		romanisation	character			
1a	affirmatory final particle (equivalent to 'x is y')	yeh	也		4	(a) Drawing of a cobra-like serpent. The semantic link, if any, would have been 'affirmation of danger'. The word for serpent, shê 蛇, is certainly related to it. But this explanation is less widely accepted than the following. (b) Drawing of the female external genitalia, the vulva. The semantic link, if any, would have been 'gate of being', hence 'affirmation of Being', and all lesser affirmations of qualities and attributes contained therein. This explanation was never challenged throughout Chinese history in spite of centuries of Confucian prudery.
1b	affirmatory verb-noun, to be, is, existence	shih	是		866c	Drawing of the sun with a foot and other strokes below it, probably composing the word chêng 正, 'correct, straight, fair and square'; not illusory. Thus 'that which exists under the sun'. Vision was here taken as representative of all the other means by which we collect sense-data.
2a	negative particle, 'not'	pu	不		999	Drawing of a flower-head on a stalk with two drooping leaves; thus, as regards sense, a borrowed homophone. The traditional explanation (Hsü Shen) was that it was an abstract concept symbol, i.e. a bird soaring aloft and not allowing itself to be caught.
2b	negative verb-noun, not to be, is not, non-existence	fei	非		579	Traditionally explained as the lower part of the word fei 飛, to fly, itself an old drawing of a bird; therefore two wings (or perhaps birds) back to back, i.e. not facing each other. Therefore (if Hsü was right) an abstract concept symbol.

		(*fei*, to fly)				
3	different	*i*	異		954	A frontal and linear representation of a man with arms raised protecting his head or making a gesture of respect. The latter meaning is found in bronze inscriptions; perhaps the gesture had reference to the assumed effulgence of a noble interlocutor. If the character is not purely a borrowed homophone, social difference between lords and people may thus have led to the idea of 'otherness, strangeness, difference' in general. The head itself is drawn in an unusual exaggerated way, possibly to represent a mask.
4	like, similar to	*ju*	如		94g	The 'woman' and 'mouth' radicals combined. A very early borrowed homophone or phonetic loan-word. No archaic significance.
5	if	*jo*	若		777	A person kneeling, perhaps gathering plants. Some have thought that submissiveness is implied, hence 'to be harmonious, to concur, complaisant', hence, by further extension, 'granted that, if . . .'. But it is more likely to be purely a borrowed homophone.
6	change, permutation	*i*	易		850	Drawing of a lizard, the meaning being derived either from colour-changes (cf. the chameleon), or rapid shifts of position.
7	change, especially gradual change, and change of form	*pien*	變		1780	Apparently not found in bone or bronze inscriptions, therefore of relatively late invention. The meaning of the drawing is uncertain, but it contains two hanks of silk and Hsü Shen said that it meant 'to bring into order', as in spinning or reeling. The radical, placed below, shows a hand holding a stick, signifying 'movement, action'. If the character is not purely a phonetic loan-word, it may have implied change from disorder to order.

There are, of course, a number of other words signifying the affirmative and the negative with various nuances.

Table 8 (*cont.*)

No.	Word	Modern Chinese		Ancient oracle-bone, bronze, or seal, form	K no.	Remarks
		character	romani-sation			
8	change, especially sudden change, and change of substance	化	*hua*	𠤎	19	Drawing of two knives, i.e. coins of knife-money. Currency exchange would thus have given rise to one expression of the idea of change in general. Cf. no. 27.
9	origin, first	元	*yuan*	𠑶	257	Drawing of a figure of a man in profile, with emphasis on the head, therefore 'first, beginning', the head being the most important part of the body. Since the ancient people doubtless knew that the head grows faster than other parts of the body during the embryonic life of vertebrates, and is relatively larger then than later, there may be an echo of primitive biological knowledge here.
10	cause, to rely on, following	因	*yin*	囡 囡	370	Drawing of a mat with woven pattern. Hence a basis, 'something to be relied on'; the meaning being extended from the static to the temporal. It may be noted that the same drawing occurs again in *hsiu* 宿, the resting-place for the night, a term of importance in astronomy.
11	cause, reason, fact	故	*ku*	𣪊	49i	The left-hand side of the ancient bronze graph is the radical meaning 'old, ancient'; its significance is not exactly known, but it originates from a drawing of a shield stored in an open rack. The right-hand side shows the hand holding the stick, symbolising action. The general meaning is clearly 'precedent' or 'prior action'.

No.	Meaning	Romanization	Character	Ref.	Description
12	make, do, act	*wei*	為	27	Drawing of an elephant, with a man's hand on its trunk, symbolising prehensility and dexterousness.
13	begin	*shih*	始	976*p*, *e′, g′, h′*	Drawing of a foetus (upside down) and a woman. Closely related to *thai* 胎, womb, and embryo. Hence here it probably signified a female embryo, a 'beginning of beginnings'.
14	go, move	*hsing*	行	748	Diagram of a crossroads.
15	go away, deprive, send away	*chhü*	去	642	Drawing of a rice-basket covered with a lid. A homophone borrowed for the present meaning.
16	come to, reach, attain	*chih*	至	413	An arrow hitting its target, or the ground.
17	stop	*chih*	止	961	Drawing of a human foot.
18	end, finished, exhausted	*chin*	盡	381	Drawing of a hand cleaning out a vessel with a brush.
19	true, the truth, real	*chen*	真	375	Seal form of uncertain representational significance, but (as we know from all the related derivative words) almost surely implying 'full, filled up, solid'. Hence the derived meaning of truth as opposed to 'empty, unreal'. The drawing of a full sack standing on a stool, if that is what it is, would thus be an abstract concept symbol.
20	above, to ascend, to hand up	*shang*	上	726	Geometrical pictograph.
21	below, to descend, to hand down	*hsia*	下	35	Geometrical pictograph.

Table 8. (*cont.*)

No.	Word	Modern Chinese		Ancient oracle-bone, bronze, or seal form	K no.	Remarks
		romanisation	character			
22	centre	*chung*	中		1007	A flagstaff with two pennants, one above the trapezoid or bushell (still used today on Chinese masts) and one below it.
23	region, side, square, quarter	*fang*	方		740	Drawing of a plough or ard. By extension, the ploughed area.
24	go in	*ju*	入		695	Drawing of a wedge or arrow-head.
25	come out	*chhu*	出		496	Drawing of a human foot, shown as leaving an enclosed space such as a cave or a house.
26	south	*nan*	南		650	Drawing of a musical instrument of some kind, perhaps a bell. How it acquired its eventual meaning is not known.
27	north	*pei*	北		909	Drawing of two men back to back. Presumably a borrowed homophone. Perhaps akin to no. 8.
28	west	*hsi*	西		594	Believed to be a drawing of a bird's nest, or (Ting Shan) a net to catch birds. Again presumably a borrowed homophone. But the graph looks very like a bundle.
29	east	*tung*	東		1175	Not, as the traditional explanation (Hsü Shen) had it, the rising sun seen through a tree, but rather a sack or bundle, which in some forms is shown being carried on a man's back. Unless the word is purely a borrowed homophone, it is curious that an azimuthal direction should be connected with a bundle. Hsü Chung-Shu regards *tung* as an

archaic form of *nang* 囊, bag, and *tho* 橐, bellows. This calls to mind the later Taoist expression *chhing nang* 青囊 'blue bag' for the heavens, hence the universe. Perhaps the equator and the ecliptic were the cords which tied up this bag. The *tho* was also the metallurgical bellows, with which Lao Tzu compares the universe. Moreover, Hsü Chung-Shu relates the custom still existing in colloquial speech of calling 'things' in general *tung-hsi* 東西, to this ancient *nang* – the heavens and earth and all that is therein. Ting Shan concurs, and Wu Shih-Chhang, though doubting the cosmic relevance of the usage, has noted interesting classical variants in the common speech of certain provinces.

No.	Form	Graph	Romanization	Meaning	Description
30	天	夨	*thien*	heaven	361	Drawing of a human figure with a large head. The obvious conclusion that it represents a primitive anthropomorphic deity has been drawn by many modern scholars.
31	日	⊙	*jih*	sun	404	Pictograph.
32	月	☽	*yüeh*	moon	306	Pictograph.
33	明		*ming*	bright, brightness	760	A combination of the two foregoing.
34	光		*kuang*	light	706	Drawing of a kneeling human figure, with fire on its head, perhaps a torchbearer.
35	歲		*sui*	year	346	Drawing originally symbolic of a special sacrifice, probably annual.
36	春		*chhun*	spring	463	Drawing of a plant sprouting in spring, with branches still not strong enough to support themselves.

Table 8. (*cont.*)

No.	Word	Modern Chinese romanisation	Modern Chinese character	Ancient oracle-bone, bronze, or seal, form	K no.	Remarks
37	summer	*hsia*	夏	[character]	36	Drawing of unknown significance. The right-hand top element represents a pig.
38	autumn	*chhiu*	秋	[character]	1092	Drawing of a tortoise. The character later evolved, through a series of stages now identified, into a combination of 'grain' and 'fire'.
39	winter	*tung*	冬	[character]	1002	Probably not a drawing of two pendent icicles (Hsü Shen), but of falling branches with fruit or leaves on them.
40	wind	*fêng*	風	[character]	625	Borrowed homophone from a somewhat similarly written character depicting the phoenix, or more properly speaking the peacock (*Pavo cristatus*), with its feathers displayed. The phonetic on the right of the bone form is probably, and suitably, a sail.
41	rain	*yü*	雨	[character]	100	Pictograph of raindrops.
42	snow	*hsüeh*	雪	[character]	297	Pictograph of snowflakes.
43	lightning	*tien*	電·	[character]	385	Hsü Shen thought that this was an attempt to draw something far-stretching which accompanies rain. He interpreted *shen* 申, one of the cyclical characters, as a symbol for stretching. But elsewhere he interpreted *shen* as a pictogram of lightning, and this is now to be accepted. The zigzag flash is accompanied by drops of rain. In *shen* 神, deity, divinity, the lightning graph persists as phonetic,
		(*shen*)		[character]		

44	thunder	*lei*	雷		577	indicating that the lightning was regarded by the ancient Chinese with the same awe as the thunderbolts of Zeus, Thor and Indra were by others. It is thus all the more striking that the wielder of the lightning, if originally fully personified, did not keep his personality long in the Chinese mind.
45	rainbow	*hung*	虹		1172*j*	To represent the noise of thunder there was added to the lightning-pictograph a drawing of wheels rumbling among the flashes. The round objects have also been taken to be drums, but this is less plausible.
46	life, birth	*sêng*	生		812	Drawing of a type of two-headed serpent in the heavens. The zoomorphic (rain-dragon?) aspect persists in the modern character as the 'insect' radical.
47	with, together, belonging to the same group as	*thung*	同		1176	Drawing of a plant rising out of the ground; symbolic of vegetal growth.
48	group, class, category	*lei*	類		529	Drawing of a vessel covered with its lid. Some bronze forms give the latter a handle. A vessel and its lid certainly belong together.
49	young	*shao*	少		1149	No early forms known. The semantic significance is uncertain. In some ancient texts (such as the *Shih Ching*) the word meant 'good'. The graph has 'head' and 'rice' as phonetic with 'dog'. The traditional explanation was (apparently) that dogs were dogs, although there were many breeds of dog looking rather unlike each other.

Drawing of four cereal grains, a concept symbol for paucity of grain, hence 'fewness' in general, and 'young' by extension of meaning. *(entry 49)*

Table 8. (cont.)

No.	Word	Modern Chinese romanisation	character	Ancient oracle-bone, bronze, or seal form	K no.	Remarks
50	old	lao	老		1055	A drawing of an old man leaning on a stick.
51	death	ssu	死		558	Drawing of a man kneeling beside bones or a skeleton.
52	man, human being	jen	人		388	Drawing of a male human being.
53	man	nan	男		649	Drawing of a field and a plough. The male tiller of the soil is implied.
54	woman	nü	女		94	Drawing of a female human being.
55	body	shen	身		386	Drawing of the body of a pregnant woman.
56	blood	hsüeh	血		410	Drawing of a sacrificial vessel with its contents.
57	self	chi	己		953	Borrowed homophone. The drawing probably represents the wound cord of a tethered arrow.
58	male, ancestor	tsu	祖		46	Phallus, hence phallic-shaped ancestral tablet. Originally *tsu* 且, now used only for animals. Connected with *mu* 牡, now used only for animals.
59	female, ancestor	pi	妣		566n	Female external genitalia. In another form *phin* 牝, now used only for animals.
60	ruler, duke, public, just	kung	公		1173	Said to be again the male generative organ, the *glans penis* being emphasised. The traditional view (Hsü Shen) was

that the character was a combination of *pa* 八, = *pei* 背, and *ssu* 私, i.e. 'turning the back on private interest', but this is not convincing.

No.	Meaning	Romanization	Character	Ref.	Notes
61	lines, design, pattern, ornament, a pictogram, literature, civilian, civilised	*wên*	文	475	A human figure viewed frontally, showing tattoo-marks or painted designs on the body.
62	sunny, bright, the south side of a hill, the Yang force	*yang*	陽 易	720	Traditionally (Hsü Shen), the upper part of this character is the sun; while the lower part represents slanting sun-beams (Wu Shih-Chhang). Cognate forms suggest a drawing of a man holding up a perforated jade disc, the *pi* 璧. This was not only a ritual object, but also perhaps, as will later be seen, the most ancient of Chinese astronomical instruments.
63	shady, dark, the north side of a hill, the Yin force	*yin*	陰	460, 651x, y, z	Drawing of *yün* 云 clouds, combined with *chin* 今, as phonetic, and (as in the previous word), *fou* 阜, hill.
64	metal	*chin*	金	652	Perhaps a drawing of a mine shaft with a cover or a hill above, the dots indicating lumps of ore.
65	wood	*mu*	木	1212	Pictograph of a tree.
66	water	*shui*	水	576	Pictograph of running water.
67	fire	*huo*	火	353	Pictograph of flames.
68	earth	*thu*	土	62	Drawing of the phallic-shaped altar of the god of the soil.

Table 8. (*cont.*)

No.	Word	Modern Chinese romanisation	Modern Chinese character	Ancient oracle-bone, bronze, or seal, form	K no.	Remarks
69	vapour, steam, subtle matter	*chhi*	氣		517c	Pictograph of rising vapour. The 'rice' component was a late addition.
70	way (in which Nature works, or which Society ought to follow)	*tao*	道		1048	Picture of a head (symbolising a person), *heading* somewhere on a road, hence 'way', hence 'the right way'.
71	natural pattern, the veins in jade, to cut jade according to its natural markings; principle, order, organisation	*li*	理		978d	No bone or bronze forms known. The graph given is a suggested reconstruction only. 'Field' and 'earth' are certainly the phonetic, 'jade' is the radical. The character must have been invented relatively late.
72	natural regularity, rule, law	*tsê*	則		906	(*a*) Perhaps a drawing of a knife being used to carve a set of pictures, or inscribe a code of laws, upon a ritual cauldron of bronze or iron.

73 — *tu* — measured division, limit, bound, law — **906**

(b) Alternatively, the reference may have been to the table-manners of aristocrats as the pattern or rule which others should follow. In this case the knife was an eating-knife and the pot a flesh-pot.

801

No bone or bronze forms known. Drawing of which the essential part is the hand below. The hand (and arm) was one of the most important standards of measurement in ancient times.

74 — *fa* — method, model, to model, mould, law (especially human positive law) — (old form) — **642k, l, m**

The original form of this word combined 'water' with 'to go away' and *chai*, a legendary one-horned bull or unicorn. This animal was supposed to gore the guilty party in an ordeal at law before the altar of the god of the soil. Evil was thus driven out (made to go away), if indeed the unicorn or some other animal did not afterward play the part of a scapegoat itself. As late as the Han time there are instances of offenders being sent to fight with wild beasts in arenas. The water component probably arose, not (as Hsü Shen thought) from the belief that 'the law should be as level as water', but from the lustrations, aspersions, libations, or sprinklings which accompanied the ancient ceremony. Perhaps a 'sink-or-swim' ordeal was also involved.

75 — *lü* — rule, regulation, standard musical tone — **502**

The left-hand element is half of the crossroads pictograph (no. 14 above), i.e. a street, so the semantic significance of the character was the public announcement of government orders or laws – since the right-hand element depicts a hand holding a writing-brush. Hence the meaning of standardisation. Somewhat later the character came to be connected with the musical tones of the standard pitch-pipes. No bone or bronze forms known.

Table 8. (*cont.*)

No.	Word	Modern Chinese		Ancient oracle-bone, bronze, or seal, form	K no.	Remarks
		romani-sation	character			
76	virtue, power, property, *mana*	tê	德		919k	The drawing combines the left-hand side of the crossroads pictograph (no. 14 above) with the primitive anatomical representations of the eye and the heart. The two latter certainly refer to seeing and thinking respectively. The former certainly refers to the social matrix. Hence the original meaning of this word was probably closely analogous to that of *mana* and *virtus*; the 'magnetic' power possessed by a leader of men whether priest, prophet, warrior or king, who came, saw, reflected, and conquered. The word *virtus* also first had to do with *man* in the fullest sense (*vira*, hero). Hence, by extension, the *mana* or numinous quality of certain inanimate objects. Later, the 'virtues' of herbs and stones. Or of the Tao.
77	ceremonial, *mores*, ethical social behaviour, natural law (juristic)	li	禮		597	Drawing of a ritual vessel containing two pieces of jade. Combined (especially in writing later than the bone or bronze inscriptions) with the radical meaning 'sign, signify, show, inform, deity, divinity, religious'. Some have thought that this was a drawing of the long and short divining sticks laid out. But others (Kuo Mo-Jo) believe, very justifiably, that it is a disguised form of the phallic symbol (cf. no. 68 above).

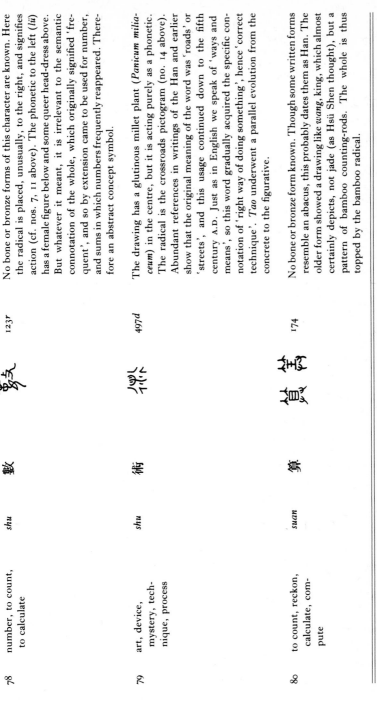

K no.	Character	Ancient forms	Reading	GSR no.	Meaning	
78	數		*shu*	123r	number, to count, to calculate	No bone or bronze forms of this character are known. Here the radical is placed, unusually, to the right, and signifies action (cf. nos. 7, 11 above). The phonetic to the left (*lü*) has a female figure below and some queer head-dress above. But whatever it meant, it is irrelevant to the semantic connotation of the whole, which originally signified 'frequent', and so by extension came to be used for number, and sums in which numbers frequently reappeared. Therefore an abstract concept symbol.
79	術		*shu*	497d	art, device, mystery, technique, process	The drawing has a glutinous millet plant (*Panicum miliaceum*) in the centre, but it is acting purely as a phonetic. The radical is the crossroads pictogram (no. 14 above). Abundant references in writings of the Han and earlier show that the original meaning of the word was 'roads' or 'streets', and this usage continued down to the fifth century A.D. Just as in English we speak of 'ways and means', so this word gradually acquired the specific connotation of 'right way of doing something', hence 'correct technique'. *Tao* underwent a parallel evolution from the concrete to the figurative.
80	算		*suan*	174	to count, reckon, calculate, compute	No bone or bronze form known. Though some written forms resemble an abacus, this probably dates them as Han. The older form showed a drawing like *wang*, king, which almost certainly depicts, not jade (as Hsü Shen thought), but a pattern of bamboo counting-rods. The whole is thus topped by the bamboo radical.

K no. from B. Kalgren's etymological dictionary *Grammatica Serica* in *Bulletin of the Museum of Far Eastern Antiquities*, Stockholm, 1940, vol. 12, p. 1. (Photographically reproduced as a separate volume, Peking, 1941).

the beginnings of science were formed just as one might expect, given the principle of the ideograph. Only two (numbers 20 and 21) can be regarded as purely geometrical symbols; the remaining seventy-eight are drawings of one kind or another. Of these one still defies analysis, at least eight are borrowed like-sounding words (homophones), and three or four concern abstract ideas, but otherwise the characters depict natural objects, the human body and its parts, and human activities. This might be expected, but what is remarkable about them is that characters concerned with technology and communication are the most numerous of all. Such a bias might change if we analysed a larger sample, but even so we can see how from everyday life ideographs were developed which could later acquire quite abstract meanings. They demonstrate, too, how technical terminology for everyday thinking and experimentation actually arose.

The School of Naturalists (Yin–Yang Chia), Tsou Yen and the origin and development of the Five-Element theory

The Five-Element theory goes back to between 350 and 270 B.C., to the time of Tsou Yen, the real founder of all Chinese scientific thought. Although he may not have been the original inventor of the theory, he certainly systematised and stabilised the ideas about it that had been in circulation for more than a century before. Something of Tsou Yen's character and prestige may be culled from the *Shih Chi* (Historical Record) (first century B.c):

'He (Tsou Yen) saw that the rulers were becoming ever more dissolute and incapable of valuing virtue... So he examined deeply into the phenomena of the increase and decrease of the Yin and the Yang, and wrote essays totalling more than 100000 words about their strange permutations, and about the cycles of the great sages from beginning to end. His sayings were vast and far-reaching, and not in accord with the accepted beliefs of the classics. First he had to examine small objects, and from these he drew conclusions about large ones, until he reached what was without limit. First he spoke about modern times, and from this went back to the time of Huang Ti. The scholars all studied his arts...

'He began by classifying China's notable mountains, great rivers and connecting valleys; its birds and beasts; the fruitfulness of its water and soils, and its rare products; and from this extended his survey to what is beyond the seas, and men are unable to observe.

'Then starting from the time of the separation of the Heavens and the Earth, and coming down, he made citations of the revolutions and transmutations of the Five Powers [Virtues], arranging them until each found its proper place and was confirmed [by history].

'Tsou Yen maintained that what the Confucians call the "Middle Kingdom" [i.e. China] holds a place in the whole world of but one part in eighty-one...

'Princes, dukes and great officials, when they first witnessed his arts, fearfully transformed themselves, but later were unable to practise them. Thus Master Tsou was highly regarded in Chhi. He travelled to Liang, where Prince Hui went out to the suburbs of the city to welcome him, and acted towards him with all the punctilio of a host towards a guest. He went to Chao, where the Prince of Phing-Yuan, walking on one side [of the road], personally brushed off the dust from his seat. He went to Yen, where Prince Chao acted as his herald, [sweeping the road with a] broom, and asked to take the seat of a disciple so as to receive his instruction.

'Here in a palace built for him at Chieh-shih, the Prince went personally to listen to his teaching... In all his travels among the feudal lords he received honours of this sort.

'Compare this with Confucius, who nearly starved to death in Chhen and Tshai, or Mencius who was surrounded with difficulties in Chhi and Liang – what a difference!

'Tsou Shih was one of the Tsou family of Chhi. He accepted the arts of Tsou Yen, and wrote essays about them. These were much appreciated by the Prince of Chhi, who gave to Shunyü Khun and all the others the title of Ta-Fu (Minister of State). He built mansions for them along a broad street, with high gates and large halls, in which they were lodged with every manifestation of respect. And the guests of the other feudal Princes and Dukes said that Chhi was able to attract all the great scholars of the world.'

This long quotation is instructive. The visits of Tsou Yen to the feudal courts are certainly historical, and he seems to have been the oldest member of Prince Hsüan's important Chi-Hsia Academy, located outside one of the gates of the capital city of Chhi; but what is especially worth noting is that, unlike the Taoists, Tsou Yen and his followers the Naturalists did not shun the life of courts and kings. Moreover, it is obvious that all the philosophers of this proto-science enjoyed great social importance and prestige, but this would hardly have been so if the 'arts' of the Naturalists were merely verbal. There must have been more to it than that.

Further examination of the *Shih Chi* gives the clue: the Naturalists' ideas carried with them an element of political dynamite which the feudal rulers were not slow to realise. Its nature can best be described in the words of Tsou Yen himself:

> 'The Five Elements dominate alternately. [Successive emperors choose the colour of their] official vestments following the directions [so that the colour may agree with the dominant element].
>
> 'Each of the Five Virtues [Elements] is followed by the one it cannot conquer. The dynasty of Shun ruled by the virtue of Earth, the Hsia dynasty ruled by the virtue of Wood, the Shang dynasty ruled by the virtue of Metal, and the Chou dynasty ruled by the virtue of Fire.
>
> 'When some new dynasty is going to arise, Heaven exhibits auspicious signs to the people. During the rise of Huang Ti [the Yellow Emperor] large earthworms and large ants appeared. He said, "This indicates that the element Earth is in the ascendant, so our colour must be yellow, and our affairs must be placed under the sign of Earth." During the rise of Yü the Great, Heaven produced plants and trees which did not wither in autumn and winter. He said, "This indicates that the element Wood is in the ascendant, so our colour must be green, and our affairs must be placed under the sign of Wood..." During the rise of the High King Wên of the Chou, Heaven exhibited fire, and many red birds holding documents written in red flocked to the altar of the dynasty. He said, "This indicates that the element Fire is in the ascendant, so our colour must be red, and our affairs must be placed under the sign of Fire. Following Fire there will come Water. Heaven will show when the time comes for the *chhi* of Water to dominate. Then the colour will have to be black, and affairs will have to be placed under the sign of Water. And that dispensation will in turn come to an end, and at the appointed time, all will revert once again to Earth. But when that time will be we do not know."'

So the Naturalists had a half-scientific, half-political doctrine with which they could frighten their feudal masters, even though the basic conception of the Five Elements (Earth, Wood, Metal, Fire and Water) was essentially naturalistic and scientific, as will shortly become clear. It was certainly Tsou Yen who extended it into the dynastic world, teaching that every ruler or ruling house reigned only 'by virtue of' one of the elements of the series. He provided, in effect, a theory of the rise and fall of ruling houses, bringing human affairs and their history under the same 'law' as the phenomena of non-human Nature. The mechanism of both was the same, an unvarying uniformity of reacting relationships between the elements which came to be known as the 'Mutual' or 'Cyclical Conquest'. All changes in human history, then, were manifestations of the same changes observable at lower 'inorganic' levels. The feudal lords seem to have become convinced of the truth of the doctrine, although they were then faced with very serious problems. On the one hand it was difficult to ascertain under what elements they might be ruling so that they might take any necessary precautions; on the other the cyclic changes of Nature were inexorable, so no ruling house could expect to go on for ever. Obviously they needed expert advice, and were willing to pay well for it.

Although the Naturalists' view was no doubt largely speculation, it seems, nevertheless, to have rested on more than that, and to have involved some practical arts as well, 'arts' that included some astronomy and calendar-making, as well as early alchemy. Tsou Yen, as we have seen, made lists of natural products – probably minerals, chemicals and plants – but his disciples went further:

> 'Moreover from first to last Sung Wu-Chi, Chêng Po-Chhiao,
> Chhung Shang and Hsienmên Kao were all people [from the
> state of] Yen who practised the method of [becoming]
> immortals by the use of magical techniques, so that their
> bodies would be etherealised and metamorphosed by some
> transmutation.'

From this it seems that they were in contact, if not identical, with the magical technicians known to have lived in the sea-board states, and who were so important at the court of the autocratic Han Emperor Wu Ti. There is evidence also of the transmission of secret writings and, probably, oral traditions too. In the *Chhien Han Shu* of A.D. 100 or thereabouts, we read:

> 'Now [the Prince of] Huai-Nan had had in his pillow [for safe
> keeping] certain writings entitled *Hung Pao Yuan Pi Shu* (The
> Secret Book of the Precious Garden). These writings told
> about divine immortals and the art of inducing spiritual beings
> to make gold, together with Tsou Yen's technique for
> prolonging life by a method of repeated [transmutation].'

Of course alchemical writings of the second century B.C. may have been fathered on Tsou Yen although never written by him – it has always been a habit of alchemists to do this – but it does seem clear that Chinese alchemy, which was older than that of any other part of the world, began among the Naturalists.

The theory of the Five Elements also comes in the *Shu Ching* (Historical Classic), a patchwork of pieces from various dates, appearing there in a Chhin interpolation that is probably about the third century B.C. and certainly not older than the time of Tsou Yen. The passage begins by saying that the doctrine of the Five Elements was part of the ninefold Great Plan, the other parts of which, except for one astronomical section on time, were concerned with human and social qualities and relations. The parts are termed 'invariable principles' (*i lun*), the character *i* being of special interest here since its original ideograph (彝) showing a ritual vessel of pork and rice, garlanded with silk and held up by two hands, appears to be a liturgical instruction, whereas here it is used in the sense of a (natural) scientific law.

The section that deals with the Five Elements is useful because it gives us some insight into the way the Naturalists saw them. It is not easy to translate, but runs in this wise:

'As for the Five Elements, the first is called Water, the second Fire, the third Wood, the fourth Metal and the fifth Earth. Water [is that quality in Nature] which we describe as soaking and descending. Fire [is that quality in Nature] which we describe as blazing and uprising. Wood [is that quality in Nature] which permits of curved surfaces or straight edges. Metal [is that quality in Nature] which can follow [the form of a mould] and then become hard. Earth [is that quality in Nature] which permits of sowing, [growth] and reaping.

'That which soaks, drips and descends causes saltiness. That which blazes, heats and rises up generates bitterness. That which permits of curved surfaces or straight edges gives sourness. That which can follow [the form of a mould] and then become hard produces acridity. That which permits of sowing, [growth] and reaping, gives rise to sweetness.'

This makes it clear that the conception of the elements was not so much of five sorts of fundamental matter as of five sorts of fundamental process, Chinese thought again concentrating on relation rather than substance. We can, indeed, construct a table showing first the element, then the natural property or process which struck the Naturalists, followed by its modern equivalent, and lastly the corresponding taste the Naturalists associated with it.

WATER	soaking, dripping, descending (dissolving?)	liquidity, fluidity, solution	saltiness
FIRE	heating, burning, ascending	heat, combustion	bitterness
WOOD	accepting form by submitting to cutting and carving instruments	solidity involving workability	sourness
METAL	accepting form by moulding when in the liquid state, and the capacity of changing this form by re-melting and re-moulding	solidity involving congealing and re-congealing (mouldability)	acridity
EARTH	producing edible vegetation	nutrivity	sweetness

On this view, the Five-Element theory was an attempt to classify the basic properties of material things, properties, that is, which would only be manifest when they were undergoing change. The 'elements' were, in fact, five powerful forces in ever-flowing cyclical motion, not passive motionless fundamental substances. In the West, the ancient Greeks had elements which did refer rather to substances, recognising three as early as the seventh century B.C. and five by about 560 B.C. They were Earth, Air, Fire, Water, with a fifth 'essence' that acted as a kind of substratum of the others. Yet even so there were some parallels with the Chinese, for the Greeks spoke of the elements warring with one another – a remarkable parallel with the Chinese Mutual Conquest theory – and even associated elements with particular gods. However, it was not until Aristotle, Tsou Yen's older contemporary, that the Greek elements became properly associated with qualities. Yet in spite of such parallels, the differences between the Chinese and the Greek views are still more striking than the similarities, and it seems unnecessary to assume any transmission of ideas between West and East.

The Naturalists seem to have looked on their elements with something of a chemical bias, though their scientific approach was modified as time passed, especially during its transmission to the Han. A key figure here was Fu Shêng, a scholar of the old state of Chhi, who was probably born soon after the death of Tsou Yen and who worked mainly between 250 and 175 B.C. He was an expert on the *Shu Ching*, which, tradition has it, he could repeat almost entirely from memory. The story may well be apocryphal; it may mean that he and the group round him re-edited it, but whatever the truth, it seems that it was from this time onwards that the Five-Element theory became incorporated into it. And then, as further scholars dealt with the theory in the years that followed, it became more political and less scientific, and by the third quarter of the first century A.D., the essentials of the theory had become surrounded by an enormous accretion of omens and portent-lore of all kinds, culminating in the theory of 'Phenomenalism', a belief that government or social irregularities would lead to dislocations of the Five-Element process on Earth and deviations from the proper course of events in the heavens. Here the proto-science of the Naturalists turned into a pseudo-science, which we shall discuss in the next chapter.

This was a crucial transformation largely due to the corruption of texts. What had happened was that by the first century A.D. Chinese culture had polarised into two schools of thought – those of the 'Old Text School' and those of the 'New Text School', the latter containing most of the scientific and pseudo-scientific thinkers. The division had arisen because of the discovery of a set of versions of the Chinese classics that differed from those previously accepted. Written in an archaic script, they appeared in 92 B.C. when Prince Kung of Lu was enlarging his palace and what was supposed to be the house of Confucius was pulled down, but later discussion made it clear that the 'archaic' texts were probably forgeries. However, the situation was complex, for although the New Text School were on stronger textual ground they accepted all the exaggerations of the pseudo-sciences, and while the Old Text School pinned their faith on false documents, they were concerned with more rational material. Generally speaking, the New Text School was dominant in the Early Han, and their opponents in the Later.

Literature on the Five-Element theory from just before and during the Han is prolific, tedious and fanciful, for the theory became increasingly connected with prophecy and divination. For instance, in discussing the dominance of each element over a particular season, the *Kuan Tzu* book says:

'When we see the cyclical sign *ping-tzu* arrive, the element Fire begins its reign. If the emperor now takes hurried and hasty measures, epidemics will be caused by drought, plants will die, and the people perish. After seventy-two days this period is over.

'When we see the cyclical sign *wu-tzu* arrive, the element Earth begins its reign. If the emperor now builds palaces and constructs pavilions, his life will be in danger, and if city walls are built [at this time] his ministers will die. [For the people should not be taken away from their harvesting.] After seventy-two days this period is over.'

Again the important Confucian natural philosopher Tung Chung-Shu, writing about 135 B.C., said:

'Heaven has five elements, first Wood, second Fire, third Earth, fourth Metal, and fifth Water. Wood comes first in the cycle of the five elements and Water comes last, Earth being in the middle. This is the order which Heaven has made. Wood produces Fire, Fire produces Earth [i.e. ashes], Earth produces Metal [i.e. as ores], Metal produces Water [either because molten metal was considered aqueous or, more probably, because of the ritual practice of collecting dew on metal mirrors left out at night], and Water produces Wood [for woody plants require water]...As transmitters they are fathers, as receivers they are sons. There is an unvarying dependence of the sons on the fathers, and a direction from the fathers to the sons. Such is the Tao of Heaven.

'The fact that definite propositions can be made about them means that sage men can get to know them, and thereby increase their own loving-kindness and decrease their severity, lay stress on the nourishing of life, and take care about the funeral offices for the dead, in this way being in keeping with Heaven's ordinances. Thus as a son welcomes the completion of his years [of nurture], so Fire delights in Wood, and as [the time comes when] the son buries his father, so [the time comes when] Water conquers Metal...

'Thus Wood has its place in the east and has authority over the *chhi* of spring. Fire has its place in the south, and has authority over the *chhi* of summer. Metal has its place in the west, and has authority over the *chhi* of autumn. Water has its place in the north, and has authority over the *chhi* of winter. This being so, Wood takes charge of life-giving, and Metal of death-dealing; Fire of heat, Water of cold. Men have no choice but to go by this succession; officials have no choice but to operate according to these powers. For such are the calculations of Heaven.'

In later centuries the cyclical recurrence of the elements became much stylised, and in due course there were even twelve phases to correspond with the twelve months of the year, using each of the Five Elements in turn. These elaborations were much used in fate calculations.

The Five-Element theories were incorporated into Han political thinking but by then the proto-scientific aspect of Tsou Yen's thought was emphatically rejected by the conventional and orthodox Confucians. How emphatically we can see in the *Yen Thieh Lun* (Discourses on Salt and Iron) written by Huan Kuan about 80 B.C. and supposedly a verbatim account of a conference held the year before between officials and Confucian scholars:

> 'The Lord Grand Secretary said: "Master Tsou was sick of the later Confucians and Mohists who did not understand the vastness of Heaven and Earth, and the Tao of the universe, broad and bright. Knowing only one part, they thought they could talk about all nine parts; knowing only one corner of the world they thought they understood the whole of it. They thought they could determine heights without a water-level, and tell the difference between straight lines and arcs without using squares and compasses. But Tsou Yen was able to make inferences about the cycles of the great sages from beginning to end, giving examples [from history]..."
>
> 'The scholars answered, "Yao appointed Yü to be Minister of Works and to control the waters and the land. Following the natural course of the mountains he marked out the heights with wooden posts, and delimited the Nine Provinces. But Tsou Yen was no sage; with strange and deceptive teachings he enchanted the six feudal princes, and so got them to accept his ideas. This is what the *Chhun Chhiu* calls 'the bewilderment of the feudal kings by one common fellow'. Confucius said 'People do not know how to manage human affairs; how should they know about the affairs of the gods and spirits?'... Therefore gentlemen [*chün-tzu*] should have nothing to do with things which are of no practical use. What is not concerned with government matters he should not investigate."'

This passage is important not only because it shows the anti-scientific view of the Confucians, but also because it reveals how strongly the ruling house of Chhin was influenced by the Naturalists. Fortunately the alchemical and pharmaceutical components were not lost by their Confucian rejection, since they became absorbed into the Taoist complex of ideas.

The stabilised Five-Element theory

We are now in a position to consider the Five-Element theory as it was finalised in Han times, and so handed down to all later ages. Two aspects merit special attention: the Enumeration Orders and the Symbolic Correlations.

The *Enumeration Orders* are orders in which the five elements were named in various ancient and mediaeval presentations of the subject. They were far from always being the same, but the four most important were as follows:

i	The Cosmogonic Order	$w\ F\ W\ M\ E$
ii	The Mutual Production Order	$W\ F\ E\ M\ w$
iii	The Mutual Conquest Order	$W\ M\ F\ w\ E$
iv	The 'Modern' Order	$M\ W\ w\ F\ E$

where w is Water, W Wood, F Fire, M Metal and E Earth.

The Cosmogonic Order was the order in which the elements were supposed to have come into being. It begins with Water, thus echoing the recurrent emphasis in Chinese writings on Water as the primeval element. The Mutual Production Order was the order in which the elements were supposed to give rise to one another; it gave the seasons in correct order, beginning with Wood for the Spring and Water for Winter (Earth corresponding to a month situated in between summer and autumn).

The Mutual Conquest Order described the series in which each element was supposed to conquer its predecessor. It was, in a sense, the most venerable order since it was the one associated with the teaching of Tsou Yen himself. It was based on a logical sequence of ideas that had their basis in everyday scientific facts: for instance that Wood conquers Earth because, presumably, when in the form of a spade, it can dig up earth. Again, Metal conquers Wood since it can cut and carve it; Fire conquers Metal for it can melt or even vaporise it; Water conquers Fire because it can extinguish it; and, finally, Earth conquers Water because it can dam it and contain it – a very natural metaphor for people to whom irrigation and hydraulic engineering were so important. This order was also considered significant from the political point of view; it was put forward as an explanation of the course of history, with the implication that it would continue to apply in the future and was, therefore, useful for prediction. Lastly, there was the 'Modern' Order, of obscure significance. However, it is this order that has come down to us in Chinese colloquial speech, where everyone still learns 'Metal, Wood, Water, Fire, Earth', even in nursery rhymes.

There are two interesting secondary principles involved in the Enumeration Order, principles that concern Rate of Change; these are the Principles of Control and Masking. The Principle of Control was derived solely from the Mutual Conquest Order, and according to it a given process of conquest is said to be controlled by the element that conquers the conqueror. For example, Metal conquers Wood but Fire controls the process; Fire conquers Metal but Water controls the process, and so on. The idea was used in fate-calculations, but nevertheless the Chinese were following in it perfectly logical paths of thought which, in our own time, have been found applicable in numerous fields of experimental science. For instance, in the ecological balance of animal species, the abundance of various forms that prey on one another in a sequence depends on their sizes and habits; increases in the numbers of a particular bird, for example, will indirectly benefit the population of greenfly because of the thinning effect on ('conquest of') ladybirds, which eat the greenfly but are themselves eaten ('controlled') by the birds.

The Chinese principle of control could of course be criticised on the grounds that as a cyclic process it meant that nothing could ever happen because every element would always inhibit another. But the Chinese never supposed that all elements were effectively present everywhere at the same time, so the criticism is purely formal. And when related to the Mutual Production Order (W, F, E, M, w) as well as the Mutual Conquest Order (W, M, F, w, E), it followed that the controlling element is always the one produced by the conquered element, and the system could carry on. For example, Wood conquers Earth in a process controlled by Metal, but Metal is a product of Earth, so there is bound to be a feedback. This idea could have social consequences as when the Confucians adopted it to prove that a son had the right to take revenge on the enemy of his father. But its primary connotations were scientific. The idea that something acting on something else destroys it, and yet, in so doing, is itself affected so as to bring about its own change or destruction was known in ancient alchemy. It is also familiar to the chemist of today, where the chemical reaction which comes to a stop because of an accumulation of reaction products is one example. Biochemistry has even more telling ones, e.g. one of the contractile proteins of our muscles, myosin, is itself the enzyme adenosine-triphosphatase that breaks down the substance bringing energy to it.

The second principle, the Principle of Masking, depends on both the Mutual Production and the Mutual Conquest Orders. It refers to the masking of a process of change by some other process which creates more of the material than is being destroyed, or makes it faster. Thus, Wood destroys (conquers) Earth, but Fire masks the process, since Fire will destroy Wood and make Earth (ash) at a greater rate than Wood can destroy Earth. Again, as with the Principle of Control, there are modern biological and ecological examples of this, as when the action of large carnivores in devouring lemmings in Norway is masked by other factors that enormously increase the lemming population.

It is important to note that in both these principles there lurks a strong quantitative element: the conclusions depend on quantities, speeds and rates. This arose, perhaps, from questions raised simply by the Enumeration Orders themselves, for we should realise that the early Chinese thinkers were not satisfied with them. This can be seen, for instance, in the *Mo Ching* where a fragment of the criticisms of the Naturalists by the late Mohists is preserved:

'C The Five Elements do *not* perpetually overcome one another.
CS The five are Metal, Water, Earth, Wood, and Fire. Quite apart [from any cycle] Fire naturally melts Metal, if there is enough Fire. Or Metal may pulverise a burning Fire to cinders, if there is enough Metal. Metal will store Water [but does not produce it]. Fire attaches itself to Wood [but is not produced from it].'

This attack on Tsou Yen's Mutual Conquest theory may be a retort to his supercilious attitude to the logical studies of the Mo-Ming schools, but is interesting all the same as a demonstration of the quantitative approach in Mohist scientific thinking.

The symbolic correlations

The Five Elements gradually came to be associated with every conceivable category of things in the universe that it was possible to classify in fives. Table 9 sets forth some of these, but it must be realised that this is only a selection.

If one divides the correlations into groups, as has been done in drawing up the table, a careful analysis makes it seem likely that the different groups were compiled by different groups of scholars. The astronomical group, for instance, may go back as early as the ninth century B.C., though it seems to show definite evidence of the hand of the great astronomer Kan Tê of the fourth century B.C., and probably the astrologer Kan Chung-Kho as well, who was of the same family but three centuries later. Then there were the Naturalist groups associated with Tsou Yen – of which the Yin–Yang group and the group of human psycho-physical functions are given here – and, finally, two groups of

Table 9. *The symbolic correlations*

Elements hsing 行	Seasons shih 時	Cardinal points fang 方	Tastes wei 味	Smells chhou 臭	Stems (denary cyclical signs) kan 干	Branches (duodenary cyclical signs) and the animals pertaining to them chih 支	Numbers shu 數
WOOD	spring	east	sour	goatish	chia i 甲乙	yin 寅 (tiger) and mao 卯 (hare)	8
FIRE	summer	south	bitter	burning	ping ting 丙丁	wu 午 (horse) and ssu 巳 (serpent)	7
EARTH	—a	centre	sweet	fragrant	wu chi 戊己	hsü 戌 (dog), chhou 丑 (ox), wei 未 (sheep) and chhen 辰 (dragon)	5
METAL	autumn	west	acrid	rank	kêng hsin 庚辛	yu 酉 (cock) and shen 申 (monkey)	9
WATER	winter	north	salt	rotten	jen kuei 壬癸	hai 亥 (boar) and tzu 子 (rat)	6

Elements hsing 行	Musical notes yin 音	Hsiu hsiu 宿 (mansions)	Star-palaces kung 宮	Heavenly bodies chhen 辰	Planets hsing 星	Weather chhi 氣	States kuo 國
WOOD	chio 角	1–7	Azure Dragon	stars	Jupiter	wind	Chhi
FIRE	chih 徵	22–28	Vermilion Bird	sun	Mars	heat	Chhu
EARTH	kung 宮	—	Yellow Dragon	earth	Saturn	thunder	Chou
METAL	shang 商	15–21	White Tiger	hsiu constellations	Venus	cold	Chhin
WATER	yü 羽	8–14	Sombre Warrior	moon	Mercury	rain	Yen

Elements hsing 行	Rulers[b] ti 帝	Yin-Yang 陰陽	Human psycho-physical functions shih 事	Styles of government chêng 政	Ministries pu 部	Colours ssu 色	Instruments chhi 器
WOOD	Yü the Great [Hsia]	Yin in Yang or lesser Yang	deameanour	relaxed	Agriculture	green	compasses
FIRE	Wên Wang [Chou]	Yang or greater Yang	vision	enlightened	War	red	weights and measures
EARTH	Huang Ti [pre-dyn.]	Equal balance	thought	careful	the Capital	yellow	plumblines
METAL	Thang the Victorious [Shang]	Yang in Yin or lesser Yin	speech	energetic	Justice	white	T-squares
WATER	Chhin Shih Huang Ti [Chhin]	Yin or greater Yin	hearing	quiet	Works	black	balances

Elements hsing 行	Classes of living animals chhung 蟲	Domestic animals shêng 牲	'Grains' ku 穀	Sacrifices ssu 祀	Viscera tsang 臟	Parts of the body thi 體	Sense-organs kuan 官	Affective states chih 志
WOOD	scaly (fishes)	sheep	wheat	inner door	spleen	muscles	eye	anger
FIRE	feathered (birds)	fowl	beans	hearth	lungs	pulse (blood)	tongue	joy
EARTH	naked (man)	ox	panicled millet	inner court	heart	flesh	mouth	desire
METAL	hairy (mammals)	dog	hemp	outer door	kidney	skin and hair	nose	sorrow
WATER	shell-covered (invertebrates)	pig	millet	well	liver	bones (marrow)	ear	fear

ª The sixth month was sometimes supposed to be under the sign of Earth.

ᵇ There are many variants of this list; the names given above are those which appear in the fragment from Tsou himself, see Needham, vol. 2, p. 238, adding that of the First (Chhin) Emperor who believed his sway to be under the sign of water.

particular scientific interest, one primarily agricultural and one mainly medical. A striking point about the agricultural groups is that they contain no reference to rice in their lists of grains, although it does appear in a medical group (not given here); presumably then the agricultural correlations arose in northern China, or at an earlier date. Actually the medical philosophers had from very early times a sixfold series complementing or replacing the fivefold one of the scientific thinkers. There were thus six, not five, internal Yang organs of the body and six Yin ones, and many other examples of this classification can be found in medicine. Perhaps it originated from Babylonia where counting in sixties, not tens, was dominant; and perhaps the fivefold order was more indigenously Chinese.

As we might imagine, these correlations met with criticism, sometimes severe, because they led to many absurdities, as pointed out in the first century A.D. by Wang Chhung:

> 'The [cyclical] sign *yin* corresponds to Wood, and its proper animal is the tiger. *Hsü* corresponds to Earth, and its animal is the dog. *Chhou* and *wei* likewise correspond to Earth, *chhou* having as animal the ox, and *wei* having the sheep. Now Wood conquers Earth, therefore the tiger overcomes the dog, ox, and sheep. Again, *hai* goes with Water, its animal being the boar. *Ssu* goes with fire, having the serpent as its animal. *Tzu* also signifies Water, its animal being the rat. *Wu,* conversely, goes with Fire, and its animal manifestation is the horse. Now Water conquers Fire, therefore the boar devours the serpent, and horses, if they eat rats [are injured by] a swelling of their bellies. [So run the usual arguments.]
>
> 'However when we go into the matter more thoroughly, we find that in fact it very often happens that animals do not overpower one another as they ought to do on these theories. The horse is connected with *wu* (Fire), the rat with *tzu* (Water). If Water really conquers Fire, [it would be much more convincing if] rats normally attacked horses and drove them away. Then the cock is connected with *yu* (Metal) and the hare with *mao* (Wood). If metal really conquers wood, why do cocks not devour hares?'

This attack by Wang Chhung was part of the Chinese sceptical tradition, which will be discussed in the next chapter, but it has been quoted here because it is not untypical of the kind of strictures that appeared on the correlations as well as on the whole Five-Element theory. Yet in spite of such criticisms, it seems that in the beginning these correlations were helpful to scientific thought in China. They were certainly no worse than the Greek theory of the elements that dominated European mediaeval thinking, and it was only when they became over-elaborate and fanciful, too far removed from the observation of Nature, that they were positively harmful. As an example of their positive use, it is worth quoting from the *Mêng Chhi Pi Than* (Dream Pool Essays) of Shen Kua. Shen Kua, whose book appeared in A.D. 1086, was one of the most wide-ranging thinkers that China produced in any age, and his use of the theory is therefore of considerable interest:

> 'In the Chhien Shan district of Hsinchow there is a bitter spring which forms a rivulet at the bottom of a gorge. When its water is heated it becomes *tan fan* [bitter alum, literally "gall-alum" – probably impure copper sulphate]. When this is heated it gives copper. If this "alum" is heated for a long time in an iron pan, the pan is changed to copper. Thus Water can be transformed into Metal – an extraordinary change of substance.
>
> 'According to the (*Huang Ti Nei Ching*) *Su Wên* [the medical classic] there are five elements in the sky, and five elements on the Earth. The *chhi* of Earth, when in the sky, is moisture. Earth [we know] produces metal and stone [as ores in the mountains], and here we see that Water can also produce metal and stone. These instances are therefore proofs that the principles of the *Su Wên* are right.'

Of course Shen Kua had no clear understanding of chemical change since like everyone else at the time he was dominated by the Five-Element theory. In Europe at the same time and much later too, down to the seventeenth century, the chemical reaction that Shen Kua describes was taken as evidence of transmutation. In fact his description shows that he was a fine observer, and his account is probably the first record in any language of the precipitation of metallic copper by iron, with the consequent formation of iron sulphate.

Numerology and scientific thinking

Before we can leave the Five-Element system, two important examples of Chinese naturalistic thinking must be mentioned. Both are contained in the *Ta Tai Li Chi* (Record of Rites of the Elder Tai), a compilation made between A.D. 85 and 105, although the quotations given probably date from the second century B.C. The first runs:

> 'Sanchü Li asked Tsêng Tzu saying, "It is said Heaven is round and Earth square, is that really so?"' . . .

'Tsêng Tzu said, "That to which Heaven gives birth has its head on the upper side; that to which Earth gives birth has its head on the under side. The former is called round, the latter is called square. If Heaven were really round and the Earth really square the four corners of the Earth would not be properly covered. Come nearer and I will tell you what I learnt from the Master [Confucius]. He said that the Tao of Heaven was round and that of the Earth square. The square is dark and the round bright. The bright radiates *chhi*, therefore there is light outside it. The dark imbibes *chhi*, therefore there is light within it. Thus it is that Fire and the Sun have an external brightness, while Metal and Water have an internal brightness. That which irradiates is active, that which imbibes radiation is reactive. Thus the Yang is active and the Yin reactive.

'"The seminal essence [*ching*] of the Yang is called *shen*. The germinal essence of the Yin is called *ling*. The *shen* and *ling* (vital forces) are the root of all living creatures; and the ancestors of [such high developments as] rites and music, human-heartedness and righteousness; and the makers of good and evil, as well as of social order and disorder.

'"When the Yin and Yang keep precisely to their proper positions, then there is quiet and peace...

'"Hairy animals acquire their coats before coming into the world, feathered ones similarly first acquire their feathers. Both are born of the power of Yang. Animals with carapaces and scales on their bodies likewise come into the world with them; they are born by the power of Yin. Man alone comes naked into the world; [this is because] he has the [balanced] essences of both Yang and Yin.

'"The essence [or most representative example] of hairy animals is the unicorn, that of feathered ones is the phoenix [or pheasant]; that of the carapace-animals is the tortoise, and that of the scaly ones is the dragon. That of the naked ones is the Sage."'

This passage, which acts as an introduction to the next, is significant especially for the remark that the sage is the chief representative of the naked animals; this really is a supreme example of the fact that Chinese thought refused to separate man from Nature, or individual man from social man. But is also contains the view, which could not be bettered even by modern evolutionists, that the basic forces seen at work in the lowest creatures are the same as those which at higher levels will develop the highest manifestations of human social and ethical life. The second passage is almost entirely biological:

> 'The Master said, "[The Principle of] Change has brought into existence men, birds, animals and all the varieties of creeping things, some living solitary, some in pairs, some flying and some running on the ground. And no one knows how things seem to each of them. And he alone who profoundly scrutinises the virtue of the Tao can grasp their basis and their origin.
>
> '"Heaven is 1, Earth is 2, Man is 3. 3 × 3 makes 9. 9 × 9 makes 81. 1 governs the Sun. The Sun's number is 10. Therefore Man is born in the tenth month of development.
>
> '"8 × 9 makes 72. Here an even number follows after an odd one. Odd numbers govern time. Time governs the Moon. The Moon governs the horse. Therefore the horse has a gestation period of 11 months.
>
> '"7 × 9 makes 63. 3 governs the Great Bear [the Plough or Northern Dipper]. This constellation governs the dog. Therefore the dog is born after only 3 months.". . .
>
> 'Now birds and fishes are born under the sign of the Yin, but they belong to the Yang. This is why birds and fishes both lay eggs. Fishes swim in the waters, birds fly among the clouds. But in winter, the swallows and starlings go down into the sea and change into mussels.
>
> 'The habits of the various classes of animals are very different. Thus silkworms eat but do not drink, while cicadas drink but do not eat, and ephemeral gnats and flies do neither. Animals with scales and carapaces eat during the summer, and in winter hibernate. Animals with beaks [birds] have 8 openings of the body and lay eggs. Animals which masticate [mammals] have 9 openings of the body and nourish their young in wombs.'

Here we see that the Naturalists, or whoever it was who wrote it, were not only close observers of Nature, but also that their observations were fitted into a framework of number-mysticism. Traces of this mysticism were already evident in the table of symbolic connections (Table 9, pp. 154–5), and it appears to have exerted a fascination for a very long time, for although it began in the third century B.C., or even a little before, it was still active as late as the twelfth century A.D. This we shall have to take into account in our assessment of Chinese scientific ideas.

The theory of the Two Fundamental Forces

So far, more has been said about the Five Elements and their symbolic correlations than about the two fundamental forces of the Yin and Yang. This is merely because we know more about the Five Elements than the Two Forces, which do not appear in the surviving fragments of Tsou Yen though his school was called the Yin–Yang Chia, and in later books discussion of them was generally credited to him. Nevertheless, there can be little doubt that the philosophical use of the terms began at the start of the fourth century B.C., and that passages in older texts which use them are later interpolations.

Some facts about Yin and Yang are, however, clear. We know, for instance, that the Chinese characters for Yin and Yang are connected with darkness and light. Yin involves graphs for hill (-shadows) and clouds, and the Yang character has slanting sun-rays or a flag fluttering in the sunshine (although the latter may represent someone holding a perforated jade disc, which was the symbol of Heaven, and probably the oldest of all astronomical instruments). These correspond well with the way in which the terms are used, for example, in the *Shih Ching* (Book of Odes), a collection of ancient folksongs. Here Yin evokes the idea of cold and cloud, of rain, of femaleness, of that which is inside, dark like the underground chambers where ice is kept for summer use. Yang, on the other hand, evokes the idea of sunshine and warmth, of spring and summer months, of maleness and brightness. Yin and Yang also had more factual meanings: Yin the shady side of a mountain or valley, Yang the sunny side.

As philosophical terms their explicit use appears in the appendix of the third-century B.C. classic, the *I Ching* (Book of Changes), which will be discussed in more detail later. Here there are sayings, the sense of which is that there are only two fundamental forces or operations in the universe, now one dominating, now the other, in a wave-like succession. And there are other early mentions. The fourth-century B.C. *Mo Tzu* book, for instance, has two references, one where it is said that every living creature partakes of the nature of Heaven and Earth and the harmony of the Yin and the Yang, the other where the sage-kings are said to have brought the Yin and the Yang, the rain and the dew, in timely season. Then in that other fourth-century work, the *Tao Tê Ching* (Canon of Virtue of the Tao), it is said that living creatures are surrounded by Yin and envelop Yang, and that the harmony of their life processes depends on the harmony of the two forces.

When we move forward to the second century B.C., the time of the Han Confucians, we come across some more specific remarks. In the words of Tung Chung-Shu:

'Heaven has Yin and Yang, so has man. When the Yin *chhi* of Heaven and Earth begins [to dominate], the Yin *chhi* of man responds by taking the lead also. Or if the Yin *chhi* of man begins to advance, the Yin *chhi* of Heaven and Earth must by rights respond to it by rising also. Their Tao is one. Those who are clear about this [know that] if the rain is to come, then the Yin must be activated and its influence set to work. If the rain is to stop, then the Yang must be activated and its influence set to work.'

We shall shortly see that the *I Ching* contains some vitally important material giving us a deeper insight into the ideas of Chinese scientific thought, but since the Yin and the Yang are so basic a part of the text and its numerous appendices, it will be best to consider something of this aspect here. The *I Ching* contains a series of 64 symbolic hexagrams, each of which is composed of six lines, whole or broken, corresponding to the Yang and the Yin. Each hexagram is primarily Yin or primarily Yang, and by a judicious arrangement it was found possible to derive all the 64 in such a way as to produce alternating Yin and Yang. Figure 23 gives an example of this, showing how the Yang splits into two, then four, then eight, and so on until 64 alternations exist. And the process need not, of course, stop there; the 64 could be extended to 128, and so *ad infinitum*. The Yin and Yang components never become completely separated, but at each stage, in any given fragment, only one is manifested. Now this has considerable scientific interest because this splitting and re-splitting of two factors, with one dominant and one recessive, has parallels in modern scientific thinking, e.g. in genetics. So here once again we have a parallel to what was said about the supposed interactions of the Five Elements – they lead to paths of thought which only in our own time have been seen to have a valid application to Nature. In brief, some elements of the structure of the world as modern science sees it are foreshadowed in the speculation of the early Chinese philosophers.

'Associative' thinking and its significance

We have now reached a stage when we can see that the scientific ideas of the Chinese involved two fundamental principles, the Two Forces and the Five Elements. Basically the Two Forces, the Yin and Yang, were derived from the negative and positive projections of Man's own sexual experience, while the Five Elements were believed to lie behind every substance and every process. With these Five Elements were associated or correlated everything in the universe susceptible of a fivefold arrangement, although, since not everything could be grouped in this way, there was a larger region comprising everything else that

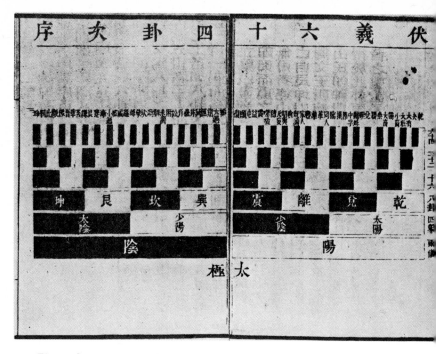

Fig. 23. Segregation Table of the symbols of the Book of Changes (*Fu-Hsi Liu-shih-ssu Kua Tzhu Hsü*), from Chu Hsi's *Chou I Pên I Thu Shuo* (twelfth century A.D.). Yin and Yang separate, but each contains half of its opposite in a 'recessive' state, as is seen when the second division occurs. There is no logical end to the process but here it is not followed beyond the stage of the 64 hexagrams.

was classifiable but which would only go into some other order (fours, nines, twenty-eights, etc.). This wider approach gave rise to another aspect of Chinese thought – number-mysticism, of which the main purpose was to relate the various numerical categories with one another.

Most European observers have written off this number-mysticism as pure superstition, and have claimed that it prevented the rise of true scientific thinking in China. Some modern Chinese scientists have also been inclined to take the same view, but at least they had the excuse that they had to deal with many thousands of traditional Chinese scholars who, unschooled in any modern scientific view, still imagined that the ancient thought-system of China was a viable alternative. But we are not concerned with this last problem, with the modernisation of Chinese society, which is quite capable of modernising itself; we have to discover whether the ancient and traditional thought-system was merely superstition or simply a variety of primitive thought; whether, perhaps, it contained something characteristic of the civilisation which produced it, and if it did, whether it contributed something positive to other civilisations.

A study of primitive magic shows that this seems to have operated on the basis of two 'laws', the 'law of similarity' according to which like produces like, and the 'law of contagion' whereby things that have once been in contact, but are not so any longer, still continue to act upon one another. Yet these are just the kind of 'laws' that lie behind the Chinese correlations or associations, which fit in completely with the ideas of similarity and contagion. Immediately, then, we have a clue to the motive lying behind the compilation of their immense correlative lists, it was magical; a fact that need not disturb us since we have already noted that in early times it was magic that nourished science and that probably the earliest scientists were magicians. After all, though magic arises in a wide variety of ways from the mystical life from which it draws strength, it then mingles with the life of the ordinary man to serve him. It tends to the concrete, the factual; it deals with reality. Indeed, magic works in the same sense as techniques work, as operations in chemistry, or in industry; it is an art of doing things. But there is little need to labour the point further, for it has already been stressed in the chapter on Taoism; what we see now, though, is that, because of this practical emphasis of magic, the symbolic correlations of the Chinese were just what a magician would need to practise his art: they brought order into things and did this at a time when no one could know what would bring success and what would not in any magical or experimental procedure. There had to be some guide to choosing the right conditions, and the correlations provided it. If one were experimenting or doing magic with water then it seemed logical, say, not to wear red, the colour of fire. Such associations might be more intuitive than anything else, but at the time what else could they be?

Modern scholars have called this kind of mental approach 'associative thinking' or 'co-ordinative thinking'. It is a system that works by association and intuition, and it has its own logic and its own laws of cause and effect. It is not just superstition, but a thought-form perfectly reasonable by its own standards, though, of course, it differs from the type of thinking characteristic of modern science, where the emphasis is on external causes. It does not classify its ideas in a series of ranks but side by side in a pattern. Things influence one another not by mechanical causes but by a kind of induction effect.

In the Chinese thought with which we are dealing, the key words are *Order* and *Pattern*; or one might almost say there is only one key-word, *Organism*, for certainly the symbolic correlations, the correspondences, the hexagrams of the *I Ching*, all formed part of one gigantic whole. Things behaved in particular ways not necessarily because of the prior actions of other things, but primarily because their position in the ever-changing cyclical universe was such that they were endowed with intrinsic natures which made such behaviour natural for them. If they did not behave in those particular ways they would lose their positions, and their relations to other things (which made them what they were) would alter and would turn them into something other than themselves. Their existence depended on the whole world-organism, and they reacted on one another by a kind of mysterious resonance.

Nowhere is this better expressed than by Tung Chung-Shu in the *Chhun Chhiu Fan Lu* (String of Pearls on the Spring and Autumn Annals) of the second century B.C. In a chapter entitled 'Things of the Same Genus Energise Each Other' we read:

'If water is poured on level ground it will avoid the parts which are dry and move towards those that are wet. If [two] identical pieces of firewood are exposed to the fire, the latter will avoid the damp and ignite the dry one. All things reject what is different [to themselves] and follow what is akin. Thus it is that if [two] *chhi* are similar, they will coalesce; if notes correspond they resonate. [The experimental proof of this is extraordinarily clear. Try tuning musical instruments.] The *kung* note or the *shang* note struck upon one lute will be answered by the *kung* or the *shang* notes from other stringed instruments. They sound by themselves. There is nothing miraculous, but the Five Notes being in relation; they are what they are according to the Numbers [whereby the world is constructed].

'[Similarly] lovely things summon others among the class of lovely things; repulsive things summon others among the class of repulsive things. This arises from the complementary way in which a thing of the same class responds – as for instance if a horse whinnies another horse whinnies in answer, and if a cow lows, another cow lows in response.

'When a great ruler is about to arise auspicious omens first appear; when a ruler is about to be destroyed, there are baleful ones beforehand. Things indeed summon each other, like to like, a dragon bringing rain, a fan driving away heat...

'It is not only the two *chhi* of the Yin and Yang which advance and retreat, according to their categories. Even the origins of the varied fortunes, good and bad, of men, behave in the same way. There is no happening that does not depend for its beginning upon something prior, to which it responds because [it belongs to the same] category, [*lei*].'

The classification which Tung Chung-Shu uses is the capacity of various things in the universe to fit into a fivefold, or some other, numerical grouping. And it is interesting that he takes the acoustic resonance of stringed instruments as an example of this, for to those who knew nothing of sound waves it must have seemed very convincing, proving his point that things in the cosmos that belonged to the same class resonated with, or energised, one another. He did not, of course, take the very primitive view that anything could affect anything else: his relationships were part of a closely knit universe with selective effects. Indeed, to Tung Chung-Shu, and to his successors, causation was something very special, since it acted in a sort of stratified pattern, not at random. Nothing was uncaused, but nothing was caused mechanically. The organism of the universe was such that everything fitted into its place and acted according to an eternal dramatic cycle; if anything missed its cue it would cease to exist. But nothing ever did fail in this way. And here, as elsewhere in Chinese thought, the text shows that the regularity of natural processes is conceived of, not as a government of law, but of mutual adaptations to community life. Not only in human relationships but throughout the world of Nature, there was give and take, a kind of mutual courtesy rather than strife among inanimate powers and processes. Solutions were found by compromise.

If all this expresses something deeply true about the Chinese world-picture – and there is much to make us believe that it does – then the fivefold correlations are an abstract chart of the whole thought-system, and the scholars of Han and later times were far from being stuck in the mud of 'primitive thought'. In true primitive thought anything can be the cause of anything else: everything is credible; nothing is impossible or absurd. If a steamship with one funnel more than usual calls at a small seaport and an epidemic follows, the appearance of the steamship is just as likely as anything else to be regarded as the cause. But once things are categorised, as they were in the fivefold system, then anything can in no way be the cause of anything else.

With these thoughts in mind, we are driven to the conclusion that there are two ways of advancing from primitive truth. One was the way taken by some of the Greeks: to refine the ideas of causation in such a way that one ended up with a mechanical explanation of the universe, just as Democritus did with his atoms. The other way is to systematise the universe of things and events into a structural pattern which conditioned all the mutual influences of its different parts. On the Greek world-view, if a particle of matter occupied a particular place at a particular time, it was because another particle had pushed it there. On the other view, the particle's behaviour was governed by the fact that it was taking its place in a 'field of force' alongside other particles that are similarly responsive: causation here is not 'responsive' but 'environmental'.

The Greek Democritean approach may have been a necessary prelude to modern science, but that does not mean that the criticism of the Chinese view as mere superstition is correct; far from it. The idea that things belonging to the same class resonated with, or energised, each other was echoed also in Greece. Aristotle, for instance, claimed that there were three kinds of 'motion': movement in space was explained by asserting that like attracts like; growth by asserting that like nourishes like; change of quality by saying that like affects like. This view, and others akin to it, echo the earliest Greek philosophers who spoke of 'love' and 'hatred' in natural phenomena. But the point to be emphasised here is that while Greek thought as a whole moved away from these views towards concepts of mechanical cause and effect, Chinese thought developed the organic concept. It is a mistake, and a serious one, to think of this Chinese outlook on Nature as essentially primitive. It was a precisely ordered universe, not governed either by the fiat of a supreme creator-lawgiver nor by the inexorable clashes of atoms, but by a harmony of wills, spontaneous but ordered in patterns, rather like the dancers in a country dance, none of whom are bound by law or pushed by the others, but who co-operate voluntarily. If the Moon stood in a certain constellation of stars at a certain time, it did so not because anyone ordered it to do so, nor yet because it was obeying some regularity, some isolatable cause, which could be expressed mathematically. It did so because it was the nature of the pattern of the universal organism that it should do so, and for no other reason. Looking back down the long avenues of time, we see the universe of Newton at the end of the Democritean view, but we do not find an emptiness at the other; instead there is the modern 'philosophy of organism' that stems from the twentieth-century mathematician and philosopher A. N. Whitehead, and of which we shall have more to say shortly.

The contrast between the two views of the universe, the traditional Chinese and that generally accepted by modern science, comes out very clearly in their use of numbers. A large amount of creditable mathamatics was, of course, done in China, as we shall see, but the point at issue is the Chinese use of number-mysticism or numerology in connection with their associative thinking. Numerology – the kind of thing epitomised in the nineteenth-century fancies associated with the Great Pyramid, where the lengths and intersections of passages are taken as providing the dates of future events – is utterly distasteful to the modern scientific mind. In China too it seems to have contributed little of scientific value, but equally important, it does not appear to have had any really bad effect either; indeed, it can be claimed that even the most exaggerated 'numerical' correlations of the Five Elements were valid in their way. Certainly they played their part in the development of Chinese scientific thinking, just as in Europe extravagances like the legal trials of animals for misdemeanours foreshadowed the conception of Laws of Nature.

Time and space were also looked on differently in the East and the West. For the ancient Chinese time was not a purely abstract quantity, but divided into separate seasons, each with their own subdivisions. Nevertheless there was continuity too, because time flowed in one direction only, and in China there was never any disposition to adopt the cyclical recurrent time of Indian philosophers, even though their beliefs were known. Space was not something abstractedly uniform, extending in all directions, but was divided into separate regions – south, north, east, west and centre – each connected with time and with the Five Elements into 'correspondences'. The east was indissolubly connected with the spring and with Wood, the south with summer and Fire, and so on. This compartmented world was quite similar to that of mediaeval Europe before Galileo and Newton had extended geometrical space and universal gravitation to the entire cosmos.

For the ancient Chinese, things were connected rather than caused, as Tung Chung-Shu put it in the second century B.C.:

> 'The constant course of Nature is that things in opposition to each other cannot both arise simultaneously. The Yin and Yang [for example] move parallel to each other, but not along the same road; they meet one another, and each in turn operates as controller. Such is their pattern.'

The universe is a vast organism, with now one component, now another, taking the lead at any one time, with all the parts co-operating in a mutual service which is perfect freedom.

In such a system as this, causality is not like a chain of events, but rather like what the modern biologist calls the 'endocrine orchestra' of mammals where, though all the endocrine glands work, it is not easy to find which element is taking the lead at any one time. And we should be clear about it; modern science needs concepts like this when considering questions like the higher nervous centres of mammals and even Man himself. But leaving modern science aside, it is clear that the concept of causality where the idea of succession was subordinated to that of interdependence dominated Chinese thinking.

Element theories and experimental science in China and Western Europe

The direct effects of the Chinese outlook on Western science may be more marked than is usually suspected. In this century there is a movement to rectify the mechanical universe of Newton by a better understanding of the meaning of natural organisation. It represents a trend that is beginning to run through all modern investigations into the methods of enquiry used in the natural sciences and the world-picture formed by them. In biology it has already put an end to the sterile arguments about whether an organism is driven by a mechanism or a vital principle inherent in living things. Yet if this movement is traced back to its beginnings it is found to have originated not in the twentieth century but in the seventeenth, its progenitor being the philosopher and mathematician Gottfried Leibniz. Co-discoverer with Newton of the infinitesimal calculus, founder of modern symbolic logic, he wanted to devise a 'science of motion that unites matter with forms'. In the end this led him to put forward the idea that the ultimate reality in the universe is a 'monad', a kind of psycho-physical entity. Monads are indestructible, have no causal relation with other monads but contain within themselves the principle of change. At the creation of the universe the monads were perfectly synchronised with one another in pre-established harmony, and each now spontaneously mirrors all changing reality without being affected by other monads. This was indeed a philosophy that seems to contain echoes of the Chinese view of Nature, especially so when we find Leibniz writing to a contemporary that, in the growth of a plant from a seed, the present state of a substance must involve its future states and *vice versa*. It is important, therefore, to realise that Leibniz had, in fact, studied Chinese ideas in Jesuit translations of works of the twelfth-century Neo-Confucian school of Chu Hsi (see p. 231). In fact this introduction to Chinese thinking probably helped Leibniz materially in the task of working out his philosophical scheme. And since the modern Western philosophy of organism owes most of its origin to Leibniz, it owes something to Chinese scientific thinking as well.

However, there is one aspect of the European concept of the universe as an organism that goes back before Leibniz, and this is the doctrine of the microcosm and the macrocosm; it contained many similarities to Chinese thinking about a cosmic pattern, but it never dominated thought to the same degree. The Western doctrine had two aspects: on the one hand there was a detailed correspondence, part for part, between the universe and the body of Man; on the other there were correspondences between the human body and the state. The doctrine seems first to have been prominent in the fourth century B.C. with Plato and Aristotle, Aristotle being the first to use the word microcosm. In his *Physics* he wrote:

> 'If this can happen in the living being, what hinders it from happening also in the All? For if it happens in the little world [microcosm], [it happens] also in the great.'

The Greek Stoic philosophers continued what Plato had begun, and most of them argued that the world was an animate and rational being. This, of course, invited detailed correspondence between Man and Nature, and by the first century A.D. we find the Roman philosopher and statesman Seneca clearly stating that Nature was like the body of Man, water-courses corresponding to veins, geological substances to flesh, earthquakes to convulsions, and so on.

This view persisted throughout late antiquity and the Middle Ages in Europe, even though some Christian fathers opposed it for a time. It permeated into Islam, appearing as a doctrine of the 'Brethren of Sincerity' at Basra in the tenth century, and even after the Renaissance and the arrival of modern science, some sixteenth- and seventeenth-century nature philosophers still held the belief. The Paracelsian physician and mystic Robert Fludd, for instance, set up opposites as follows:

> Heat – Movement – Light – Dilatation – Attenuation
> Cold – Inertia – Darkness – Contraction – Inspissation

an example that might have been taken from any Chinese exponent of Yin and Yang. The famous friar Giordano Bruno, who was burned at the stake in 1600, also regarded the universe as an organism, and spoke of sexual intercourse between the sun and the earth whereby all living creatures were brought into being. But these views seem to have derived from Pythagoras, the Greek mystic and mathematician of the fifth century B.C., rather than from Plato, whatever Chinese influences there may have been.

This kind of outlook is also to be found in Jewish thought, and indeed was so widespread that one is tempted to seek a common origin, going back before Pythagoras and before the Chinese Naturalists to some source that gave the germ of the idea to both Eastern and Western civilisations. Such an origin could, perhaps, have lain in Babylonia, the civilisation of the valleys of the Euphrates and the Tigris (Mesopotamia, now part of Iraq). If little support for this is found in the Babylonian baked-clay texts, it is possible that the origin lay not in texts but in practice, and in particular the methods used for divination. There was a widespread practice of foretelling the future from the whole or part of a sacrificial animal; the Shang Chinese, as we have seen, used the shells of tortoises or the shoulder-blades of oxen and deer; the Etruscans and later the Romans used the liver, and that was just what the Babylonians had done earlier still. The idea behind such a practice was simply the theory that the heavens or the parts of an animal could be divided into areas, and that the key to the future was to be found in the 'signs' in one or another of these: the animal, in fact, acted as a microcosm of the universe. Space and time were divided into separate 'parcels', prefiguring all later scientific divisions of space and time, and within the spatial sphere the small and the great – the microcosm and the macrocosm – mirrored each other.

European and Chinese views may have had a single origin, and they certainly display some similarities, but there were also some fundamental differences. In Europe, primitive organic naturalism was accompanied by a minor counterpart, the State microcosm–macrocosm analogy, and both were subject to the characteristic European schizophrenia – the need to think either of material atoms or theological spiritualism. God the creator had always to be the prime mover behind the machine: the animal organism might be projected on to the universe, but belief in a personal god or gods meant that it always had to have a 'guiding principle'. This was a path which the Chinese emphatically did *not* take. To them, the parts of a living body or of the universe could account for the observed phenomena by a kind of will: co-operation of the component parts was spontaneous, even involuntary, and this alone was sufficient. Thus there were two traditions of the universe as an organism, and both went their separate ways. Not until the seventeenth century, when Leibniz, an heir of the European microcosm–macrocosm concept, was in communication with the Jesuits in Peking, did something of the Chinese flavour reach Western Europe and begin to exert its influence.

We are now in a position to ask whether the Five-Element theory was or was not a hindrance to the advance of natural science in China. Certainly, as we have seen, the theory could lead to absurdities, but these were no worse than the European absurdities of astrology and bodily humours that were coupled with the Western theory of elements. Looking back now, we can see that the Five-Element and Yin–Yang system was not altogether unscientific. For example, the elements corresponded rather well to what we might call today five fundamental states of matter. One could think of Water as implying all liquid, and Fire all gaseous, states; similarly Metal could cover all metals and semi-metals, and Earth all our earth elements, while Wood could stand for the whole realm of the carbon compounds, that is, organic chemistry. At any rate we should always remember that the word 'element' as a translation for *hsing* was never happy, because it has too material a connotation, and the sense of movement was implicit in the character from the beginning. Yet 'processes' or 'phases' will not do either, because the idea of matter is not entirely absent. And if anyone is tempted to mock at the persistence of these ideas in China, they should remember that in seventeenth-century Europe the protagonists of the new science had to fight a bitter action against those who wanted still to hold fast to the old Greek concepts of the universe as enshrined in the works of Aristotle. Yet, in their day, Aristotle's ideas had been in the forefront of advanced thought; it was only as time passed that they became outdated, and finally had to be thrown overboard at the Reformation and the Renaissance. And so it was with the Chinese: the Five-Element and Yin–Yang theories were on the whole helpful to the development of scientific thinking; the only trouble was that they went on too long: China had no Reformation and no Renaissance.

But when all is said and done, it may well turn out that in Western Europe the new science was still in debt to the microcosm–macrocosm concept at least. William Harvey's discovery of the circulation of the blood was not based purely on the mechanical analogy of the heart as a pump, but also on the 'correlative' analogy of the Sun and the meteorological process of water circulation on Earth: the Sun in the macrocosm had, after all, been looked on by Bruno as the equivalent of the heart in the microcosm Man. Here the parallel with China is very close, because alongside the cosmic correspondence the idea of the heart as a pump occurred there contemporaneously with Harvey. In general, of course, correlative thinking and the universe analogy in China could not survive the 'new, or experimental, philosophy' of the West; experiment, inductive reasoning from the particular to the general, and the mathematical approach to all natural phenomena, superseded all earlier forms of scientific theorising. The old idea of a cosmos of different spaces was driven out by the application of the uniform space of Greek geometry to the entire universe. But now, after a lapse of time, with a new phase in the study of biology and the growing appreciation of

Table 10. *Significances of the trigrams in the Book of Changes*

Explanation

Col. 1: The assemblage of lines which form the *kua*.

Col. 2: Romanised name of the *kua*.

Col. 3: Chinese character of the *kua*.

Col. 4a: 'Sex' of the *kua*.

Col. 4b: Associated position in a 'family' (from ch. 10 of App. 8, the *Shuo Kua*).

Col. 5: Associated animal (taken mostly from ch. 8 of App. 8, but with other information added).

Col. 6: Associated natural object or 'emblem' (from ch. 11 of App. 8). This list is important since the hexagrams shown in Table 11 are usually described in these terms. For example, *kua* no. 39, *Chien*, consists of *Khan* (trigram no. 4) over *Kên* (trigram no. 5), i.e.

$$\frac{\text{fresh-water (lake)}}{\text{mountain}}$$

Col. 7: Associated element (the five elements here have to cover eight *kua*). The list, which betrays the association of most of the appendices with the School of Naturalists, comes from ch. 11 of App. 8.

Col. 8: Associated compass-point, according to the 'more ancient' *hsien-thien*[1] ('prior to Heaven') or *Fu-Hsi*[2] system (to be discussed in a later volume).

Col. 9: Associated compass-point, according to the 'later' *hou-thien*[3] ('posterior to Heaven') or *Wên Wang*[4] system, as given in ch. 5 of App. 8 (to be discussed in a later volume).

Col. 10: Associated season.

Col. 11: Associated time of day or night.

Col. 12: Associated type of human being (from ch. 11 of App. 8).

Col. 13: Associated colour (from ch. 11 of App. 8).

Col. 14: Associated part of the human body (from ch. 9 of App. 8).

Col. 15: Primary concept or 'virtue' of the *kua* (taken mostly from ch. 7 of App. 8).

Col. 16: Secondary abstract concept of the *kua*.

[1] 先天   [2] 伏羲   [3] 後天   [4] 文王

1 kua	2	3	4a	4b	5	6	7	8	9	10	11	12	13	14	15	16
1 ☰	*Chhien*	乾	♂	father 陽	dragon, horse	heaven	metal	S	NW	late autumn	early night	king	deep red	head	Being, strength, force, roundness, expansiveness	*Donator*
2 ☷	*Khun*	坤	♀	mother 陰	mare, ox	earth	earth	N	SW	late summer, early autumn	afternoon	people	black	abdomen	Docility, nourishment of being, squareness, form, concretion	*Receptor*
3 ☳	*Chen*	震	♂	eldest son	galloping horse, or flying dragon	thunder	wood	NE	E	spring	morning	young men	dark yellow	foot	Movement, speed, roads, legumes and young green bamboo sprouts	*Stimulation, excitation*
4 ☵	*Khan*	坎	♂	second son	pig	moon and fresh water (lakes)	water	W	N	mid-winter	midnight	thieves	blood-red	ear	Danger, precipitousness, curving things, wheels, mental abnormality, abysses	*Flowing motion* (especially of water)
5 ☶	*Kên*	艮	♂	youngest son	dog, rat, and large-billed birds	mountain	wood	NW	NE	early spring	early morning	gate keepers	—	hand and finger	Passes, gates, fruits, seeds	*Maintenance of stationary position*
6 ☴	*Sun*	巽	♀	eldest daughter	hen	wind	wood	SW	SE	late spring, early summer	morning	merchants	white	thigh	Slow steady work, growth of woods, vegetative force, mercantile talent	*Penetration, mildness, continuous operation*
7 ☲	*Li*	離	♀	second daughter	pheasant, toad, crab, snail, tortoise	lightning (and sun)	fire	E	S	summer	midday	amazons	—	eye	Weapons, dry trees, drought, brightnesses, catching adherence of fire and light	*Deflagration, adherence*
8 ☱	*Tui*	兌	♀	youngest daughter (concubine)	sheep	sea and sea water	water and metal	SE	W	mid-autumn	evening	enchantresses	—	mouth and tongue	Reflections and mirror-images, passing away	*Serenity, joy*

Table 11. *Significances of the hexagrams in the Book of Changes*

Explanation

Col. 1: The assemblage of lines which form the *kua*.

Col. 2: Romanised name of the *kua*.

Col. 3: Chinese character of the *kua*. It is thought that all the names derive from those characters which occurred most frequently in the prognostications from the *kua*.

Col. 4: Characterisation of the *kua* according to its two component trigrams named by their associated natural objects or 'emblems', e.g. no. 7, *E/Fw*, earth over fresh-water; or no. 21, *L/T*, lightning over thunder.

Col. 5: One or two of the more common lexicographical meanings of the character which constitutes the name of the *kua*.

Col. 6: Concrete or social significance of the *kua*. These meanings are derived mostly from the *Ching* text itself, and from the *Thuan Chuan* commentary.

Col. 7: Abstract significance of the *kua*. These meanings represent what the *kua* came to stand for from the Han dynasty onwards, and indicate the conceptual use made of them by proto-scientific and scientific minds throughout medieval times and indeed down to the end of the tradition.

Notes will be found on pp. 180, 181.

1 *kua*	2	3	4	5	6	7
1	*Chhien*	乾	H/H	heaven, paternal, dry, male	Heaven, king, father, etc., ordering, controlling	*Donator*
2	*Khun*	坤	E/E	earth, material	Earth, people, mother, etc., supporting, containing, docile, subordinate	*Receptor*
3	*Chun*	屯	Fw/T	sprout	Initial difficulties, 'contre-démarrage'[a]	*Factors slowing the onset of a process*
4	*Mêng*	蒙	M/Fw	cover	Youthful inexperience[b]	*Early stages of development*
5	*Hsü*	需	Fw/H	need, procrastinate	Dilatory policy[c]	*Stopping, waiting*

6		*Sung*	訟	H/Fw	litigation	Strife, contention at law[d]	*Opposition of processes*
7		*Shih*	師	E/Fw	army, general, teacher	Military affairs[e]	*Organised action*
8		*Pi*	比	Fw/E	assemble	Union, concord	*Coherence*
9		*Hsiao Hsü*	小畜	WW/H	to rear	Creative force modified by mildness, taming	*Lesser inhibition*
10		*Li*	履	H/Sw	shoe, to tread	Hazardous success attained by circumspect behaviour, treading delicately	*Slow advance*
11		*T'hai*	泰	E/H	prosperous	Geniality of spring, peace (in the Sung came to mean one of the progressive world periods)	*Upward progress*
12		*Phi*	否	H/E	bad	Beginning of autumn (in the Sung came to mean one of the retrogressive world periods)	*Stagnation, or retrogression*
13		*Thung Jen*	同人	H/L	lit. people together	Union, community	*State of aggregation*
14		*Ta Yu*	大有	L/H	lit. great having	Abundance of possessions, opulence	*Greater abundance*
15		*Chhien*	謙	E/M	humility	Hidden wealth, modesty	*Highness in lowness*
16		*Yü*	豫	T/E	pleased	Harmonious excitement, enthusiasm, satisfaction	*Inspiration*
17		*Sui*	隨	Sw/T	follow	Following	*Succession*

Table 11 (*cont.*)

No.	Name	Chinese	Code	Meaning	Description	
18	Ku	蠱	M/WW	virulent poison	Troublesome work in a decaying society^f	*Corruption*
19	Lin	臨	E/Sw	approach	Approach of authority	*Approach*
20	Kuan	觀	WW/E	to look / soundofvoices courts, criminal law	Contemplation, looking for omens,^g letting influence radiate	*View, vision*
21	Shih Ho	噬嗑	L/T	gnawing;^h sound of voices	Crowds, markets, courts and criminal law	*Biting and burning through*
22	Pi	賁	M/L	bright	Ornamental	*Ornament, pattern*
23	Po	剝	M/E	to peel, flay	Falling, overthrowing, collapse, like a house held together only by its roof (*kua* pictographic)	*Disaggregation, dispersion*
24	Fu	復	E/T	return	Year's turning-point	*Return*
25	Wu Wang	无妄	H/T	not reckless, not false^i	No recklessness, no insincerity, not guilt, yet difficulties	*Unexpectedness*
26	Ta Hsü	大畜	M/H	to rear	Creative force suppressed by something stationary and heavy	*Greater inhibition*
27	I	頤	M/T	jaws	Mouth (which the *kua* shows in pictographic form)	*Nutrition*
28	Ta Kuo	大過	Sw/WW	to overstep	Large excess, strangeness not necessarily unfavourable^j	*Greater top-heaviness*

No.	Figure	Name	Character	Trigrams	Meaning	Description	
29		Khan	坎	Fw/Fw	pit[k]	The edge of the ravine, danger and the reaction to it; below, the torrent of water	*Flowing motion*
30		Li	離	L/L	separate, apart	The meshes of a net (*kua* pictographic), catching adherence of fire and light	*Deflagration, adherence*
31		Hsien	咸	Sw/M	all (but here used for *kan*)[l]	Mutual influence, interweaving, wooing	*Reaction*
32		Hêng	恆	T/WW	constant[m]	Perseverance	*Duration*
33		Thun	遯	H/M	to hide oneself, conceal[n]	Withdrawal, retreat	*Regression (further advanced than no. 12)*
34		Ta Chuang	大壯	T/H	great strength[o]	Great strength	*Great power*
35		Chin	晉	L/E	to rise, advance[p]	Advance in feudal rank	*Rapid advance*
36		Ming I	明夷	E/L	intelligence repressed[q]	Lack of appreciation of the services of a good official	*Darkening, extinction of light*
37		Chia Jen	家人	WW/L	family people	Members of a family or household	*Relation*
38		Khuei	睽	L/Sw	separated	Division and alienation	*Opposition*
39		Chien	蹇	Fw/M	lame[r]	Lameness, inhibition	*Retardation*
40		Chieh	解	T/Fw	dissection, analysis	Unravelling	*Disaggregation, liberation*

Table 11 (*cont.*)

No.	Name	Character	Code	Gloss	Description	Summary
41	Sun	損	M/Sw	spoil, hurt, subtract	Removal of excess, payment of taxes	*Diminution*
42	I	益	WW/T	benefit	Increase of resources, addition	*Increase, addition*
43	Kuai	夬	Sw/H	fork, settled, decision	Breakthrough, release of strain, 'détente'	*Eruption*
44	Kou	姤	H/WW	copulation	Advance to casual encounter, meeting, intercourse	*Reaction, fusion*
45	Tshui	萃	Sw/E	thicket, congregate	Process of collection, consolidation of people around a good ruler	*Condensation, conglomeration*
46	Shêng	升	E/WW	to rise	Career of a good official	*Ascent*
47	Khun	困	Sw/Fw	surrounded, distressed	Straitened, distress, bewilderment	*Enclosure, exhaustion*
48	Ching	井	Fw/WW	a well	Dependableness	*Source*
49	Ko	革	Sw/L	skins	Moulting of skins, hence change	*Revolution*
50	Ting	鼎	L/WW	tripod cauldron	Nourishment (of talents) (*kua* alleged pictographic)	*Vessel*
51	Chen	震	T/T	quake, rock, thunder	Moving exciting power	*Excitation*

52	Kên	艮	M/M	limit^s	Stability, as of a mountain	*Immobility, maintenance of stationary position*
53	Chien	漸	WW/M	gradually tinge^t	Slow and steady advance (like chemical changes produced by soaking – dyeing, retting, lixiviating)	*Development, slow and steady advance*
54	Kuei Mei	歸妹	T/Sw	lit. returning, younger sister	Marriage^u	*Union*
55	Fêng	豐	T/L	abundance (good harvest)	Prosperity	*Lesser abundance*
56	Lü	旅	L/M	travel, travellers	Strangers, merchants	*Wandering*
57	Sun	巽	WW/WW	gentle	Penetration of wind	*Mildness, penetration*
58	Tui	兌	Sw/Sw	exchange	Sea, pleasure	*Serenity*
59	Huan	渙	WW/Fw	broad, swelling, irregular	Dispersion, alienation from good	*Dissolution*
60	Chieh	節	Fw/Sw	joints of bamboo	Term, section, regular division, regulation, meditation (on general phenomena), confinement, silence	*Regulated restriction*
61	Chung Fu	中孚	WW/Sw	lit. central; confidence	Inmost sincerity, kingly sway	*Truth*
62	Hsiao Kuo	小過	T/M	to overstep slightly	Small excess	*Lesser top heaviness*

Table 11 (*cont.*)

63	䷾	*Chi Chi*	*Fw/L*	lit. end; up to the mark	Completion, successful accomplishment	*Consummation, perfect order*
64	䷿	*Wei Chi*	*L/Fw*	lit. not quite; not quite up to the mark[v]	Position when all is not yet completed nor successfully accomplished	*Disorder, potentially capable of consummation, perfection, and order*

[a] The *Ching* text for this *kua* contains an ancient peasant omen concerning the colour of the horse on which the bride arrives at her husband's house. A marriage is one of the things which have 'initial difficulties'.

[b] All the interpretations given diverge from what was the original sense of this *kua*. It concerned the Chinese equivalent of the Golden Bough. *Mêng*[1] was anciently another name for the dodder, an epiphyte like the mistletoe. Its commoner names are *nü lo*[2] (women's net) and *thu ssu tzu*[3] (rabbit silk). The absence of roots in the dodder was a matter of great interest for the early Chinese naturalists, as we shall see later in connection with the question of action at a distance (and magnetism). Earlier still there can be no doubt that it was an important magical object. Parasitic plants have widely been considered sacred. A. Waley has shown that the phrase with which the *Ching* text opens, 'It was I who sought the *mêng* boy, it was the *mêng* boy who sought me', was simply a spell for averting the evil consequences of tampering with the holy plant. The idea of 'youthful inexperience' arose from the ancient folk custom of thinking of this plant as a boy, and of course the ultimate abstract significance is still further removed.

[c] The significances here are based on a misunderstanding, of which we shall see further examples, due to the failure of ancient scribes to add radicals to their phonetics. The *kua* should be not *hsü*[4] but *ju*,[5] meaning some kind of creeping insect, concerning which there are five peasant-omens in the text.

[d] The contention is thought to have been about the division of war booty, for there is an omen in the text about prisoners laughing.

[e] The text has a bird-omen indicating that a parley will be successful.

[f] Here the significances built up on the basis of the *ku*-poison were all somewhat misleading; what the ancient text had to do with was the taking of omens by observations of the behaviour of maggots in the flesh of animals sacrificed to ancestral spirits.

[g] The text refers to observations of whether sacrificial animals advance or retreat, and to observations of inspired utterances of boys undergoing initiation. Waley even suggests that the original meaning of *tung*,[6] boy, was 'one who is ceremonially beaten', arguing from old forms of the character. The 'weird ditties of children' were still being attended to when Ricci gave his account of Chinese customs about A.D. 1600.

[h] The basis of the text here is omens from objects found when eating food; all the rest is wide-ranging derivation.

[i] This is all a misunderstanding. *Wu-wang* was a single word, probably meaning a figure tied to a bull which was driven away from the village as a scapegoat. This is one of Waley's most beautiful identifications.

[1] 蒙   [2] 女蘿   [3] 兔絲子   [4] 需   [5] 蠕   [6] 童

j The text contains a willow-omen.

k Pit is right; all the rest is thought-arabesques of late scholars. The ancient ceremony referred to is that of sacrificing to the moon in a pit.

l As we saw above, *hsien*[1] has lost its heart radical, and ought to be *kan*.[2] The omen basis is that of tinglings in the limbs.

m Here the concept of duration refers to what Waley calls the 'stabilising process' of ancient magicians. When once a favourable omen had been obtained it was necessary to perform a rite stabilising it (e.g. burying or locking up of objects). Since the character *hêng*[3] in its ancient form consists of the moon between two lines, the suggestion is that it was originally a rite performed at the first appearance of the new moon and directed to making a favourable condition of affairs last all through that lunar month. The lines may represent magic lines drawn round the omen in question, such as circumperambulations of new settlements. Such a view throws quite a new light on *Lun Yü*, where Confucius makes his famous remark, 'The people of the south have a saying, "It takes *hêng*[3] [usually translated as perseverance] to make even a soothsayer or a medicine-man [physician]." It's quite true. If you do not stabilise your virtue, disgrace will overtake you.' All this ends up in a perfectly abstract concept of duration.

n Here *thun*[4] is a mistake for *thun*,[5] has nothing to do with hiding, and refers to omens from movements of young pigs.

o This originates from omens about rams getting stuck in bushes.

p *Chin*[6] should be *chin*,[7] to insert. The text therefore probably had to do with magic increasing fecundity of domestic animals, insertion referring to coupling of the males and females.

q This again is a complete misunderstanding. Li Ching-Chhih showed that *ming-i* was the ancient name of a bird, and the omens concern it. The interpretations as given in the table are all later imaginations.

r As we saw above, this originates from omens about stumbling. Of course, for later scientific thinkers, a concept such as retardation was very useful.

s A complete misunderstanding. *Kên*[8] ought to be *khên*,[9] and the omen concerned the way in which rats gnawed the exposed bodies of sacrificial victims.

t Although ancient Chinese thinkers were impressed by slow chemical actions in solution, the most ancient form of this omen-text has probably nothing to do with that conception. The *kua* character can also mean 'to skim', and Li Ching-Chhih showed that it referred to the way in which the wild geese skimmed over natural objects such as rock-ledges and trees, thereby giving omens.

u Wedding-omens are at the bottom of this.

v In spite of the very high and philosophical interpretation of the final *kua*, analysis of the text of the *Ching* shows that it had once to do with omens taken from animals crossing streams. Thus Waley:

'If the little fox, when almost over the stream,
Wets its tail,
Your undertaking will completely fail.'

1 咸 2 感 3 恆 4 遯 5 豚 6 晉 7 搢 8 艮 9 齩

the nature of living organisms, the correlative approach of Leibniz, and thus of the Chinese, is having a part to play, although in modified forms.

The system of the Book of Changes

Much has already been said of the Chinese belief that things resonate with one another – the world-view of what we should now call 'action at a distance'. And we have said that the Five-Element and Yin–Yang theories were a help rather than hindrance to the development of scientific ideas in Chinese civilisation. But there was a third component of Chinese natural philosophy – the system of the *I Ching*, the Book of Changes – and of this it will not be possible to form so favourable a judgement.

Originating from what was probably a collection of peasant omen texts, and accumulating a mass of material used in divination, it ended up as an elaborate system of symbols and their explanations, without counterpart in the texts of any other civilisation. The symbols were supposed to mirror in some way all the processes of Nature, and Chinese mediaeval scientists were therefore continually tempted to rely on pseudo-explanations of natural phenomena by simply referring them to the symbol to which they might be supposed to pertain. The symbols themselves gradually developed abstract significance and came to save the necessity for further thought, their very abstraction giving a deceptive profundity.

As we have it today, the *I Ching* is a complex book. The symbols are composed of sets of lines (see Tables 10 and 11) as already mentioned under the discussion of the Two Forces. The lines of these may have been connected with counting rods, or with the long and short sticks used in divination. All possible permutations and combinations of the lines give eight trigrams or three-lined sets (Table 10), and 64 six-lined sets (hexagrams) (Table 11), all known as *kua*: and they are arranged in the book in a definite order. Each *kua* is provided with a single paragraph of explanation, traditionally attributed to King Wên of the early Chou dynasty (about 1050 B.C.); then there is a commentary, usually in six sentences, supposedly by Chou Kung (about 1020 B.C.). But this is only the main text; it is accompanied by even more complex commentaries and appendices, some traditionally (though wrongly) by Confucius himself.

The data of the book are still a matter of some uncertainty. No one now believes that King Wên or the Duke of Chou had anything to do with it, though some think it probably goes back to the sixth century B.C., while others prefer to believe that it was written in the third during the Warring States period. For our purpose it will perhaps be best to adopt the view that the basic text originated from omen compilations of the seventh and eighth centuries B.C., but did not reach its present form until the end of the Chou dynasty (third century B.C.). The commentaries and appendices would then date from the first century B.C. to the first century A.D., and would have been compiled by Chhin and Han Confucians.

The peasant omens, derived from ancient Chinese farmers, were like those to be found in every civilisation where people are in a similar stage of primitive culture: they concern inexplicable sensations and involuntary movements, unusual phenomena observed in animals and plants, and unusual meteorological and astronomical events. These were then all overlaid with the divinations to which they were supposed to have given rise. For example, if we look at *kua* 39, *Chien*, which is derived from the idea of stumbling and means 'lame', we find it is said to have the practical and social significance of lameness and inhibition, and the abstract significance of 'retardation'. The commentaries on it run:

> '[In the state indicated by] *Chien* advantage [will be found] in the south-west, and the contrary in the north-east. It will also be advantageous to meet with a great man. [In these circumstances] with firmness and correctness [there will be] good fortune'

and

> '(1) [From the] first line, divided, [we learn that] advance [on the part of its subject] will lead to [greater] difficulties, while remaining stationary will afford ground for praise.
> (5) The fifth line, undivided, shows its subject struggling with the greatest difficulties, while friends are coming to help him.'

The very ancient peasant omen enshrined in the second commentary would have been as follows: 'He who goes stumbling shall come praised: a great stumble means a friend shall come.'

If the *I Ching* had been no more than a divination text, it would merely have taken its place with a host of others of a similar kind. But the addition of detailed appendices gave it a higher ethical and cosmological status, and the more abstract the explanations became, the more the system as a whole assumed the character of a repository of concepts to which every concrete fact of Nature could be referred, and from which a pseudo-explanation could be found for every event. Furthermore, the growing abstraction of the symbols kept pace with the development of early science out of early magic. To Han scholars really trying to take a naturalistic attitude to subjects like magnetism or the tides, this must have seemed the obvious thing to do. Unfortunately it was misleading, and perhaps they would have been wiser to tie a millstone round the neck of the *I Ching* and cast it into the sea.

We can gain an idea of the kind of repository of concepts that the *I Ching* became by noting that, of the 64 *kua*, a total of 45 deal with some aspect or other of time and space, leaving only 19 *kua* that do not, *kua* like inspiration, truth, vision or unexpectedness, for example. The fact that so many *kua* can be fitted into a 'space–time' relationship indicates the extent to which Chinese thinkers after the Han managed to get the *kua* away from the extremely human significances they once had had. The *kua* can, of course, also be put into other categoories, though these only show relations between the *kua* which may well have no counterpart in Nature. Unhappily mediaeval Chinese scholars set more store by these than by anything observed in the natural world.

In assessing the significance of the *I Ching* for Chinese science, there are two other relevant questions we must ask: what, according to the appendices, did the School of Naturalists and the Han scholars think it was all about; and, second, how were the significances of the *kua* made use of in succeeding centuries by scientific writers? The answer to the first question seems to be that, as we have said, they looked on the *I Ching* as a repository of concepts, and tried to extract out of the long and short lines of the hexagrams a comprehensive symbolism containing all the basic principles of natural phenomena. Like the Taoists, they were looking for peace of mind through classification, and they found such a classification in the *I Ching*.

The second question may be answered by saying that vast use was made of them. For instance, the system was extended so that the *kua* were brought systematically into association with the motion of celestial bodies, and so with the passage of time. In fact, a complete system correlating the eight trigrams with the motions of the Sun and Moon, the day of the month and the ten 'stems' (or cyclical characters) was drawn up in the late second or early third centuries A.D. and became known as the 'method of the contained stem'. This brought further connections in its train, notably in alchemy, where it was believed that the efficacy of chemical processes depended on the exact time at which they were performed. Later still, the system of the *kua* was extended to cover acoustics and speculations in biology and medicine.

The alchemical use of the *kua* can be seen in the earliest alchemical book, the (*Chou I*) *Tshan Thung Chhi* (The Kingship of the Three), written in A.D. 142 by Wei Po-Yang, and in the comments on it made a thousand years later by the Neo-Confucian philosopher Chu Hsi. Chu

Table 12. *Inventions mentioned in the Book of Changes*

| Invention | Legendary sage | Attributed to | | Explanation |
		kua	no.	
Nets (and Textiles)	Fu-Hsi	*Li*	30	*kua* alleged pictographic
Ploughshares	Shen Nung	*I*	42	*WW/T*, both associated with wood (wooden ploughs)
Markets	Shen Nung	*Shih Ho*	21	*L* (i.e. sun)/movement on roads
Boats		*Huan*	59	wood/water
Carts		*Sui*	17	sprightliness/movement on roads
Gates		*Yü*	16	*kua* perhaps pictographic; movement/earth (walls)
Pestle and Mortar	Huang Ti, Yao and Shun	*Hsiao Kuo*	62	wood/mountain (i.e. stone)
Bow and Arrow		*Khuei*	38	*L* (i.e. sun's rays like arrows)/passing away
Houses		*Ta Chuang*	34	*T* (i.e. inclement weather)/*H* (i.e. a space)
Coffins		*Ta Kuo*	28	serenity/wood
Quipu (knotted cords, as records)		*Kuai*	43	speech/solidity (i.e. retention of things spoken)

NOTE. The explanations given above are quite arbitrary: perhaps claims that inventions stemmed from the *I Ching* were simply a device to add to its prestige.

Hsi says that Wei Po-Yang made use of the method of the contained stem to guide himself to the appropriate times for adding reagents and withdrawing products to cool or settle. He also goes on to explain that while two *kua* are concerned with apparatus, and two with chemical substances, the whole of the remaining 60 are all to do with 'fire-times' – the determination of the right moments for carrying out chemical operations (and perhaps the degree of heating to be used). A study of Wei Po-Yang's text and Chu Hsi's comments makes it clear that to them the correlations in the *I Ching* embodied a dialectical element – no state of affairs was permanent, every vanquished entity would rise again, and every prosperous force carries within it the seeds of its own destruction. Moreover, Chu Hsi gives evidence, in another commentary, that the *kua* were being fully used by alchemists in the fifth century A.D. and earlier.

Table 13. *Association of the 'kua' with the lunar and diurnal cycles in the 'Tshan Thung Chhi'*

THE CYCLE OF THE LUNAR MONTH,
Tshan Thung Chhi, chs. 4 and 18

		no.	
(A)	Chen	51	Excitation
(B)	Tui	58	Serenity (i.e. the process quietly at work)
(C)	Chhien	1	Donator (i.e. maximum of maleness, no moon)
(D)	Sun	57	Mild penetration
(E)	Kên	52	Immobility
(F)	Khun	2	Receptor (i.e. maximum of femaleness) and the cycle recommences

THE DIURNAL CYCLE, *Tshan Thung Chhi*, ch. 19

		no.	
(A)	Fu	24	Point of return (i.e. starting-point)
(B)	Lin	19	Approach
(C)	Thai	11	Progression
(D)	Ta Chuang	34	Great power (i.e. acceleration of process)
(E)	Kuai	43	Decisive breakthrough
(F)	Chhien	1	Donator (i.e. maximum of maleness, noon)
(G)	Kou	44	Reaction
(H)	Thun	33	Regression
(I)	Phi	12	Stagnation
(J)	Kuan	20	Vision (?)
(K)	Po	23	Dispersion
(L)	Khun	2	Receptor (i.e. maximum of femaleness; midnight) and the cycle recommences

No less instructive is the use of the *I Ching* in the biological field, as a quotation will show. In the *Li Hai Chi* (The Beetle and the Sea – a title based on the proverb that a beetle's eye view cannot encompass the whole sea) of Wang Khuei, probably written in the late fourteenth century, we find the following remarks about blood:

'The blood of Man and animals is always red. This is because it is Yin and belongs to watery things, which are under the aegis of the *kua Khan*. But the blood also harbours a Yang [component], and it is red too because of what it contains. The interaction of *Khan* with *Li* is what causes the motion of the *chhi* [of the blood]. Now if the blood leaves the body for too long, it turns black, and if it be heated it also turns black; this is because it tends to return to its origin.'

This is typical of Wang Khuei, who noted many strange things of biochemical interest that no one else observed, but it also shows well enough the delusive nature of the *kua* system. The colour blood-red having been arbitrarily chosen in earlier centuries for association with *Khan*, now for Wang Khuei it appears as a satisfying explanation for the red colour of the blood to say that the *kua Khan* is controlling it.

The relation of the Book of Changes to the 'administrative approach'

Joseph Needham has pointed out that the powerful hold which the *I Ching* continued to exert on Chinese minds even down to recent times is a matter of general knowledge to anyone who has lived in China. Indeed, a century ago scholars were still claiming that all the truths about electricity, or light, or heat, or any other branch of European physics, were contained in the trigrams and hexagrams. Of course this is not true today; indeed the pendulum has swung so far that the history of science in Asia is suffering from the neglect of the *I Ching* by most Chinese scholars and scientists, who look on it purely as a folly of their mediaeval age. Certainly it has to be admitted that while the Five-Element and Two-Force theories were favourable to the development of Chinese scientific thought, the elaborate symbolic system of the *I Ching* was a handicap almost from the start. A method for pigeon-holing novelty and then doing nothing more about it, it became no more than a giant filing system that led to all concepts being stylised so that they fitted in to the system without difficulty, almost in the way in some ages art forms have prevented artists from looking at Nature.

There is, however, one question that cannot be dismissed. Why did the *I Ching*, to which Europe can offer no parallel, show such longevity and persistence? Can the answer be found in the description of it as a vast filing system? Was the compelling power it had in Chinese civilisation due to it giving a view of the world basically in line with the bureaucratic social order? Did it appeal because it was an 'administrative approach' to the natural world? A moment's reflection will show that the answer may well lie here. When Chinese writers say that a *kua* 'controls' such and such a time or phenomenon, when some natural object or event is said to be 'under the aegis of' such and such

Table 14. *Association of the 'kua' with the administrative system in the* Chou Li

Pu	Associated concept	kua	Tri-gram no.	Hexa-gram no.
(1) General Administration	Heaven	*Chhien*	1	1
(2) Ministry of Education	Earth	*Khun*	2	2
(3) Ministry of Rites	Spring	*Chen*	3	51
		Kên	5	52
(4) Executive	Summer	*Sun*	6	57
		Li	7	30
(5) Ministry of Justice and Punishments	Autumn	*Tui*	8	58
(6) Ministry of Public Works	Winter	*Khan*	4	29

a *kua*, one is irresistibly reminded of the phrases familiar to all who have served in government organisations – 'a matter for your department', 'passed to you for appropriate action', etc. The *I Ching* may almost, indeed, be described as an organisation for routing ideas through the right channels to the right departments, almost a heavenly counterpart of the bureaucracy on Earth, the reflection in the natural world of the unique social order of the human civilisation that produced it. And this connection was not unconscious in Chinese thought. In the idealised administrative system elaborated by Han scholars and handed down in the *Chou Li* (Record of the Rites of Chou), each seplarate ministry is actually associated with a season, and thus directly with a *kua*.

Such considerations as these lead us to what may be the dénouement of the present chapter. Perhaps the entire system of what we may call 'associative organismic' thinking was in one sense a mirror image of Chinese bureaucratic society. Not only the tremendous filing system of the *I Ching* but also the 'symbolic correlations', where everything had its position connected by 'the proper channels' to everything else, can probably best be described as an administrative approach to Nature. Perhaps, then, we should see the Taoist teaching on the universe as an organism, and the atomic world of the Greeks, as mirrors of their different environments, the Taoists in a highly organised society where bureaucracy was dominant, the atomists in a confederation of separate city-states and individual merchant adventurers.

Greek atomism and deductive geometry are part of the foundations on which the new European science of the seventeenth century rested, just as the womb of modern capitalist society gave birth in the nineteenth century to the modern science of our immediate forefathers, a science that offered a totally mechanical and materialistic view of the universe. But since their time science has had to become still more modern to take account of an enormously vaster universe at one end of the scale and a sub-microscopically small one at the other, far transcending the range of sizes for which Newton constructed his world-picture. This, and a deepening knowledge of biology, have made it necessary to take a new look at many scientific concepts in which the organism has had a vital part to play. We now find that some kind of philosophy of organism is needed, a philosophy that, by way of Leibniz, basically stems from the bureaucratic society of ancient and mediaeval China. Such a new form of modern science does not, of course, supersede the classical atomic-Newtonian system; it is simply something rendered necessary by the fact that science today has to deal with realms of the universe that the Newtonian system did not envisage. Our conclusion, then, is that Chinese bureaucratism and their organic outlook may, after all, turn out to have been a necessary element in the formation of a perfected world-view of natural science. Yet even if this conclusion is correct, it does not justify the evil effects the *I Ching* had on Chinese scientific thinking. There remains therefore the gigantic historical paradox that, although Chinese civilisation could not of itself produce modern natural science, natural science could not perfect itself without the help of the characteristic philosophy of Chinese civilisation.

There is one postscript to be added. In his correspondence with the Jesuits in Peking in the seventeenth century, and particularly with Father Joachim Bouvet, Leibniz brought himself into close contact with the *I Ching*, and as he examined the long and short lines of the hexagrams, he came to see what seemed an astonishing fact – the lines could be identified with binary numbers, counting in a scale of two (the type of counting used in computers and electronic calculators), which he himself had invented a few years earlier in place of the usual scale of ten. The Chinese seemed to him to have discovered the system many centuries before. However, this opinion of Leibniz was not quite valid: the men who had invented the hexagrams were simply concerned with all the permutations and combinations they could make from their two basic elements, the long and the short strokes. There was no Chinese thought of binary arithmetic or even any realisation that such counting could exist. However, the fact that Leibniz believed the Chinese had had this knowledge stimulated him to develop his own system, and to design the earliest computing machines.

Thus we have one further perspective. Binary arithmetic has now assumed great importance as the basic counting method used in the circuits of all electronic computers and calculators, and is also found helpful by biologists in their analyses of the central nervous system of mammals. Thus something that was unconsciously stumbled upon in the *I Ching* hexagrams, and brought to consciousness by Leibniz, has proved of fundamental use in modern science and technology.

11

The pseudo-sciences and the sceptical tradition

Superstition flourished in China just as strongly as in every other ancient culture. Divining the future, by astrology, geomancy, physiognomy, the choice of lucky and unlucky days and the lore of spirits and demons, was part of the background of ancient and mediaeval Chinese thinking. An integral part of the ancient picture of the universe, it simply cannot be ignored if we are to get a balanced picture of Chinese science, especially since some of this superstition led, almost imperceptibly, to important discoveries and practical investigations of the natural world. After all, magic and science both involve manual oprations, so the empirical element was never missing, even from Chinese 'pseudo-science', and nor was scepticism. A sceptical approach was always a factor in Chinese thinking, just as it is for all true natural knowledge. On these two counts, then, vital elements necessary for the development of modern science were there. But a third factor was also required: the formulation of theories verifiable by the method of experiment. This the old Chinese lacked, for they were never able to pass beyond the relatively primitive and unquantifiable theories of the Five Elements and the Two Forces.

The greater part of this chapter will be devoted to the Chinese sceptical tradition, and to its greatest exponent Wang Chhung, a first-century Confucian who was nevertheless attracted by the Taoist view of Nature. But to examine his work, and to see the sceptical tradition in its proper setting, we must first turn to those very questions that aroused this scepticism – the superstitions of the pseudo-sciences.

Divination by scapulimancy and casting lots

The *shu shu*, the 'techniques of destiny', the ways of foretelling, took many forms, but all were based on the conviction that it was possible to predict the future, at least so far as the affairs of princes and the destinies of states were concerned. And they were all designed to give as unambiguous 'yes or no' answers as possible. The oldest method was scapulimancy, using red-hot metal rods to heat tortoise-shells and breast-bones, or ox and deer shoulder-blades (*scapulae*), and then interpreting the cracks that resulted. Indeed it was so old a technique that *chan*, the very word for divination, seems probably to have been derived from the pictograph of a shoulder-blade with cracks. Ox and deer *scapulae* seem to have come into use first, the tortoise-shells following later. There is some doubt about what species of tortoise was used; Chinese tradition has it that they were turtle-shells, but an examination of some of the large fragments that remain make it more likely to have been a now extinct species of land tortoise. At all events, both shoulder-blades and tortoise-shells seem to have been imported from far to the south, well outside the primary area of early Chinese civilisation.

Consultation of the cracks always gave the diviner an unambiguous 'fortunate' or 'unfortunate' answer. For the ancient Chinese this practice of resolving doubts by scapulimancy probably paid its way as a solvent for neuroses of indecision; but for us the great value of the practice lies in the help it has given for understanding of very early Chinese society and the most ancient forms of Chinese writing.

During the Chou period, that is the first millennium B.C., another divination procedure came to be adopted: the drawing of lots. Dried stalks of the Siberian milfoil (a species of yarrow) called *shih* were used (Fig. 24). This was also the same word as the technical term for casting lots (*shih*), and it had the ideograph (筮) the lower part of which (巫) is the character *wu*, meaning shaman or wizard. As mentioned in the previous chapter, it was, perhaps, these milfoil sticks that were the origin of the long and short lines of the trigrams and hexagrams found in the *I Ching*.

By and large, drawing lots was adopted for affairs of lesser importance, scapulimancy with tortoise shells for matters of greater significance, but on occasions both were used at once. In this case matters were complicated, for the two methods sometimes disagreed, and then the diviner had to face two contradictory answers. From evidence in the historical classic, *Shu Ching* (Book of Documents), which, although finished in the fourth century A.D., contains material going back to the tenth century B.C., and from other texts, we find that the way out of the impasse was for the diviner to prepare a system of relationships between the different possible answers. These relationships can most conveniently be drawn up into a table (Table 15). Whether the actions taken always followed the diviner's answer as given here we do not know, but if the last category were, it must sometimes have meant a strange alliance of superstition and democracy!

Fig. 24. A late Chhing representation of the legendary Emperor Shun and his ministers, including Yü the Great, consulting the oracles of the tortoise-shell and the milfoil. From *Shu Ching Thu Shuo*.

Table 15. *Divination methods*

Pro		Contra	
T, M	or	*T, M*	Definitely favourable or unfavourable as the case might be.
T or M		*M or T*	Milfoil valid for the immediate future, tortoise-shell valid for the further future.
T, M, P	or	*T, M, P*	Favourable or unfavourable as the case might be, in spite of the opinions of ministers and people.
T, M, m	or	*T, M, m*	Favourable or unfavourable as the case might be, in spite of the opinions of prince and people.
T, M, p	or	*T, M, p*	Favourable or unfavourable as the case might be, in spite of the opinions of prince and ministers.

KEY. T = tortoise-shell; M = milfoil; P = opinion of the Prince; m = opinion of the Ministers; p = opinion of the People.

During Han times and later, the popularity of scapulimancy decreased, but the milfoil sticks remained in general use; indeed this continued almost down to the present day. In Taoist temples, simple folk used to choose a stick from a box rattled by the attendant *Tao shih*, and were then given a fortune-telling slip of paper corresponding to the number on the stick. Very similar procedures exist in Japanese temples today often with the use of slot-machines, and if the paper contains bad tidings you must knot it to a shrub or tree in the temple garden, thus neutralising the effect.

At the time of the Warring States period (fourth and third centuries B.C.), yet another method of divination grew up: this was the random selection of the trigrams and hexagrams of the *I Ching*, and their combination and recombination. As each one stood for more or less well-defined abstract ideas and natural processes, it was not very difficult to draw conclusions about what their fortuitous juxtapositions portended. Here, as with the milfoil sticks, and indeed as with fortune-telling methods the world over, it was the unpredictable arrangement of things after they had been shuffled that served as a prophecy, since the underlying belief was that unseen powers affected the final order. To achieve the right conditions, though, the diviner had to concentrate on what was to be known; only in this way could the spirit influences control the actions of shuffling. Divination using the *I Ching* was therefore always preceded by incense-burning and prayers, after which sticks were shuffled into either two or three groups and the odd sticks counted out by cycles of eight, thus determining the complete or broken character of each of the lines of the hexagram. Here again the psycho-therapeutic value probably lay in the resolution of painful indecision.

Astrology

Chinese astrology, which has been relatively little studied, took a somewhat different course from that of Mesopotamia and Europe. There astronomers concentrated on what celestial bodies, what constellations, appeared on the horizon just after sunset or just before dawn, as also on the motions of the Sun, Moon and planets along the ecliptic; but from the earliest times in China it was the constellations around the north pole of the sky – the circumpolar stars – that were used, together with the constellations strung along the equator. The former never rise and never set, because they are always above the horizon, even though they can only be seen at night. The equatorial constellations were then used by the Chinese to divide the heavens into 28 lunar 'mansions' or sections, like the segments of an orange. Chinese astrologers therefore paid less attention than their Western counterparts to the planet that was 'in the ascendant' at any time. What mattered to them was the particular constellation in or near which a planet or the Moon happened to stand; but also they were extremely interested in conjunctions, eclipses, comets, novae and many other 'abnormal' occurrences in the heavens, as well as meteorological phenomena that seemed strange.

In ancient Chinese astrology, as in the ancient Western world, there was little concern about the fate of individuals unless they were of royal blood: prognostications of the future were always concerned with the affairs of state, the chances of war, the prospects of the harvest, and so on. In Mesopotamia itself individual horoscopes were not introduced until about 280 B.C. by the astronomer-astrologer Berossus; a 'democratisation' of astrology followed during the next century, practised mainly by exiled 'star-clerks' from Mesopotamia. Before this change, there was little to choose between Babylonian and Chinese astrology, as is shown by comparing a Mesopotamiam baked clay tablet from the library of King Ashurbanipal (seventh century B.C.) with the text of the Chinese *Shih Chi* (Historical Record) of Ssuma Chhien (90 B.C.):

(1) Tablet: If Mars is in [the name of the constellation is missing] to the left of Venus, there will be devastation in Akkad.
Shih Chi: When Ying-huo [Mars] follows Thai-Pai [Venus], the army will be alarmed and despondent. When Mars separates altogether from Venus, the army will retreat.

(2) Tablet: If Mars stands in the house of the Moon [and there is an eclipse], the king will die, and his country will become small.

> *Shih Chi:* If the Moon is eclipsed near Ta-Chio [Arcturus] this will bring hateful consequences to the Dispenser of Destinies [the ruler].

(3) Tablet: If the Northern Fish [Mercury] comes near the Great Dog [Venus], the king will be mighty and his enemies will be overwhelmed.

> *Shih Chi:* When Mercury appears in company with Venus to the east, and when they are both red and shoot forth rays, then foreign kingdoms will be vanquished and the soldiers of China will be victorious.

The evidence here is that the Chinese observed the planets and the Moon, especially some of their movements, just as the Mesopotamians did, but they adopted different constellations for the rest of their astrological predictions. This was only natural for the entire Chinese sky was mapped in constellation patterns totally different from those of the Middle East and Europe – a fact testifying to the independent rise of ancient Chinese astronomy. It is likely, then, that what passed from Mesopotamia to China in about the first millennium B.C. was nothing more than the conviction of the possibility of a system of divination using the stars.

Just as in Europe, so in China, astrology came in the end to be used for casting individual horoscopes, but it was under foreign influence and happened rather late (Fig. 25). Certainly lunar and solar influences would lead any agricultural community to see a connection between the heavens and events on Earth, but the Chinese also knew (like the Greeks)

Fig. 25. A Chinese horoscope of the fourteenth century A.D. The nineteenth of a series of 39 sample horoscopes indicating all kinds of fortunes in life; here a person who is destined to achieve fame. Favourable features of the horoscope are shown in the top right-hand box, unfavourable ones opposite on the left. Immediately underneath and at the bottom corners are shown the celestial influences governing 42 different aspects of life and health. Among them are included, besides the sun, moon and five planets (represented by their element names), Rahu and Ketu (the nodes of the moon's path), comets and vapours. The outer ring of the disc itself gives constellation names, the third gives *hsiu*, and the seventh cyclical characters. Segment significances are defined by the fifth ring. They concern, counting counterclockwise from the right (at half-past two), fate (i.e. longevity), wealth, brothers, landed property, sons, servants, marriage and women, illness, travel, official position, happiness, and bodily constitution. From *Thu Shu Chi Chhêng.*

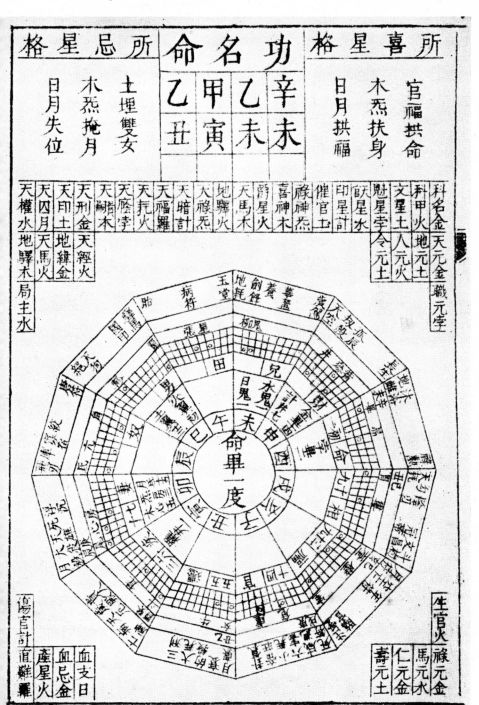

that marine animals such as sea-urchins and mussels have a reproductive cycle, the roe becoming fat and thin again in accordance with the Moon's phases, so it was not a wholly illogical step to extend to human individuals those influences which a thousand years before had been restricted only to matters of state.

Closely related to Chinese astrology were other systems of belief – the choice of lucky and unlucky days, and divining the future from the calendar. A belief in lucky and unlucky days is very old and has been traced back to Mesopotamia and Egypt; it seems to have been connected with the Moon's phases. More elaborate was the 'fate calculation' system devised in China and based on the calendar. It involved the cyclical characters known as the twelve horary 'Branches' and ten celestial 'Stems', and divination was carried out by adding the hour of an event to the day, the month and the year, thus forming four factors or 'pillars'. These pillars could then be linked with the Five Elements.

Geomancy

Besides divination based on the heavens, the Chinese also used methods that depended on the Earth, methods of geomancy or *fêng-shui* (literally 'winds and waters'). The basic idea here was that if the houses of the living and the tombs of the dead were not properly positioned, then evil effects of the most serious kind would affect the inhabitants of the houses and the descendants of those whose bodies lay in the tombs; conversely, good siting would favour their health, wealth and happiness. Adjustment to gain the desired harmony depended on local topography, since every place had features of landscape that would modify the local influence of the various *chhi* of Nature. The shapes of hills and valleys, the directions of streams and rivers, being the outcome of 'winds and waters', were the most important, but the heights and forms of buildings, the directions of roads and bridges, were also considered significant factors. Moreover, since the force and nature of invisible currents would be modified from hour to hour by the positions of the heavenly bodies, so their aspects as seen from the locality concerned had also to be considered. Siting was of vital importance; bad siting could, however, be ameliorated by digging ditches, piling up mounds or making other adjustments to alter the *fêng-shui* situation (Fig. 26).

Ideas such as these are of high antiquity, and records go back at least to the fourth century B.C., while two centuries later geomancy had become so widespread that the *Shih Chi* mentions a separate class of diviners, the *khan yü chia* (diviners by the canopy of Heaven and the chariot of the Earth). However, the real consolidation of geomancy did not take place until a century later, during the Three Kingdoms period,

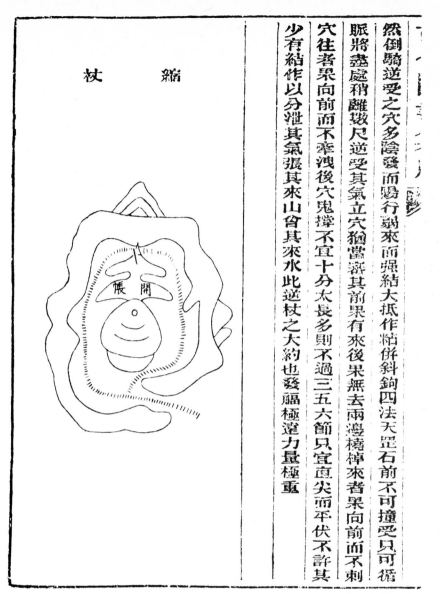

縮杖

開帳

然倒騎逆受之穴多陰發而賜行剶來而強結大抵作粘併斜鉤四法天罡石前不可撞受只可循

脈將盡處稍離數尺逆受其氣立穴猶當蔽其前果有來後果無去兩邊橈棹來者果向前而不剌

穴往者果向前而不牽洩後穴鬼撐不宜十分太長多則不過三五六節只宜直尖而平伏不許其

少有結作以分泄其氣張其來山皆其來水此逆杖之大約也發福極遠力量極重

Fig. 26. Illustration from a work on geomancy (*fêng shui*), the *Shih-erh Chang Fa* (Method of the Twelve Chang), attributed to Yang Yün-Sung of the Thang (*c.* A.D. 880). The chart shows a particular site for a tomb, towards the tip of a range of small hills separating two valleys with streams, the whole being enclosed by two further ranges of foothills. It is said that the higher these latter ranges are, the better, and that there should not be a 'tongue' or high ridge connecting the inner hills with the main massif (shown at the top as if in elevation). This kind of site is called '*so chang*' because the *chhi* of the mountain is 'condensed' around the tomb site. From *Thu Shu Chi Chhêng*.

when Kuan Lo probably wrote about it, although it is now impossible to tell how much of the *Kuan shih Ti Li Chih Mêng* (Mr Kuan's Geomagnetic Indicator) was actually his own work, or even written during his lifetime. At all events, by the time of the Three Kingdoms two currents in the Earth's surface – the Yin and Yang – had become identified with the two symbols that applied to the eastern and western quarters of the sky, the Green Dragon of Spring (east) and the White Tiger of Autumn (west). Each of these was supposed to fit in with configurations on the ground, the Green Dragon on the left and the White Tiger to the right, of any inhabited place, so as to protect it as if in the crook of an arm. Yet this was only the beginning of the complexity, since high and abrupt escarpments were considered Yang, and rounded elevations Yin, and such influences had, if at all possible, to be balanced in the selection of a site so that there were three-fifths Yang and two-fifths Yin. What is more, trigrams, hexagrams, the cycles of stems and branches, as well as the Five Elements, were all woven into the divinatory scheme and had also to be considered. As a result of all these factors, the diviners had a strong preference for tortuous and winding roads, walls and structures that seemed to fit into the landscape rather than dominate it: straight lines and geometrical layouts were avoided. In many ways *fêng-shui* was highly advantageous, as when it recommended planting trees and bamboos as windbreaks, and emphasised the value of running water close to a house site. Though in other ways, of course, entirely superstitious, it seems always to have contained a strongly aesthetic element – hence the beauty of the siting of so many farms, houses and villages throughout China.

There is now no doubt that the magnetic compass was developed for *fêng-shui*, being derived from the diviner's board, the *shih* (Fig. 27). This consisted of two layers, an upper disc corresponding to Heaven and a lower square corresponding to the Earth. The stars of the Plough (the Northern Dipper) were marked on the upper layer, and both contained the characters for the compass points. Such boards go back probably to the third century B.C., and seem quite clearly to have had to do with finding direction, even in cloudy weather, but the full story of the discovery of the magnetic compass must wait until a later volume. There is also evidence that the diviner's board had a connection with the game of chess, for its earliest use appears to have been for a form of divination carried out by casting pieces ('men') like dice on to it.

Other methods of divination

Besides the forms of divination mentioned so far, there were others based on the belief that the future of an individual could be foretold by examining his facial appearance, bodily form and so on. Physiog-

太保相宅圖

太保

Fig. 27. A late Chhing representation of the selection of a city site; the geomancer is consulting his magnetic compass. The depiction of the use of the magnetic compass in an illustration of a Chou period text is of course an anachronism. From *Shu Ching Thu Shuo*.

nomy, the study of faces, and cheiromancy (palmistry) were both used, being as widespread in China as elsewhere in the ancient world. However, in China cheiromancy had one unique effect: it led to the early discovery of finger-prints as a means of identification, earlier than in any other part of the world.

Oneiromancy, foretelling the future from dreams, was another method of divination in general use in antiquity. To what extent its practice in China involved anticipations of Freudian psycho-analysis remains for current research to reveal. Glyphomancy, predicting the future from an analysis of the written characters of names, was yet another means of divination. This was a specially Chinese method of divination that could only have arisen in a culture using an ideographic language. It may, nevertheless, have connections with Western methods of 'automatic writing' (such as 'spirit-writing' using a planchette), but that did not reach China until late Sung times (thirteenth century A.D.).

Sceptical trends in Chou and Early Han times
Although superstitious practices and fortune-telling arose very early in Chinese history, the growth of a sceptical and rationalising tradition ran them very close. Even as early as the seventh century B.C. it is said in the *Tso Chuan* (a history of the eighth to fifth centuries B.C.):

> 'Prince Li, having heard the story about the apparition of the two serpents, asked Shen Hsü about them, saying, "Do people still see apparitions of evil augury?" Shen Hsü replied, "When a man fears something, his breath [*chhi*] escaping, attracts an apparition relating to that which he fears. These apparitions have their principle in men. When men are without fault, no ominous apparitions appear. But when men throw away the rules of constant behaviour, they appear. Such is the way in which they are caused."'

Although this passage contains the age-old doctrine that moral faults give rise to natural calamities, it also expresses the idea that ghosts and apparitions are subjective, the projections of men's minds. And this is no isolated quotation; it is only one of a number of expressions of similar views. In the sixth century B.C., for instance, it was argued that a prince's health depends on his work, his journeys, his food, his joys and sorrows, etc., and not on the spirits of rivers and mountains, or on the stars. Indeed, when we come to the period of the philosophers we find many taking this kind of rationalist attitude. Thus Hsün Chhing in the third century B.C.:

'If [officials] pray for rain and get rain, why is that? I answer
there is no reason at all. If they do not pray for rain, they will
nevertheless get it. When [officials] "save the sun and the
moon from being eaten", or when they pray for rain in a
drought, or when they decide an important affair only after
divination – this is not because they think that they will in this
way get what they want, but only because it is the
conventional thing to do. The prince thinks it is the
conventional thing to do, but the people think it supernatural.
He who thinks it a matter of convention will be fortunate; he
who thinks it is supernatural will be unfortunate.'

Not all Confucians were sceptics, and during Han times Confucianism
separated into two rather sharply contrasted traditions. One arose from
the establishment of Confucianism as a state religion in the second
century B.C. and from the Taoist religion some three hundred years later,
and incorporated most of the proto-scientific and semi-magical theories
of Tsou Yen – the Yin and Yang forces, the Five Elements and a host
of divination practices. Coupled with the older Confucian ideas of
moralism, this tradition gave rise to 'phenomenalism' – the belief that
ethical or ritual faults were directly connected with cosmic irregularities.
The second tradition, on the other hand, continued the sceptical outlook, which is well epitomised in the following story of events which
happened in A.D. 46:

'When Liu Khun was prefect of Chiang-ling, his city was
devastated by fire. But he prostrated himself before it, and
immediately it went out. Later, when he became prefect of
Hung-nung, the tigers [which had previously infested the
place] swam across the Yellow River with their cubs on their
backs and migrated elsewhere. The emperor heard about these
things and wondered at them, and promoted Liu Khun to be
Chief of the Personnel Department. The emperor said to him,
"Formerly at Chiang-ling you turned back the wind and
extinguished the conflagration, and then at Hung-nung you
sent the tigers north of the river; by what virtue did you thus
manage affairs?" Liu Khun replied, "It was all pure chance."
The courtiers to left and right could not restrain their smiles
[to see a man losing such a fine opportunity of getting on in
the world]. But the emperor said, "This reply is worthy of a
really superior man! Let the annalists record it."'

As mentioned at the beginning of this chapter, the most important
figure of the sceptical tradition was Wang Chhung, one of the greatest

men China ever produced. His *Lun Hêng* (Discourses Weighed in the Balance) was finished in A.D. 83. His views were full of rationalism, and he took a thoroughly critical attitude to everything he studied. Although his conception of Nature was based on the Yin–Yang forces and the Five Elements, he denied that Heaven could be conscious, and took a naturalist world-view of which *tzu-jan*, 'spontaneity', was the watchword. The principles of maleness and femaleness seemed to him to be at the heart of Nature, since Heaven was equated with Yang and the Earth with Yin; and they gave rise, he thought, to a wave-like succession of dominance. Heaven and Man have the same Tao: if something is impossible in the Tao of Man, then it cannot come into effect in the Tao of Heaven either.

Wang Chhung took over the early ideas of condensation and rarefaction and elaborated them. Life arises from a condensation of the *chhi* of Yin and Yang:

> 'As water turns into ice, so the *chhi* crystallise to form the human body. The ice, melting, returns to water, and Man, dying, returns to the state of a spirit. It is called spirit just as melted ice resumes the name of water. When we have a Man before us we use a different name. Hence there are no proofs for the assertion that the dead possess consciousness, or that they can take a form and injure people.'

Elsewhere he says that those who study immortality and think it possible to avoid dying are doomed to failure: Man cannot live for ever any more than ice can be prevented from melting.

Wang Chhung next went on to identify the vital spirit of living things with the fiery Yang principle, and wet tissue, flesh, and bones with the aqueous Yin principle. All was well when the *chhi* of the Yin and Yang were in good order, but if the fiery Yang *chhi* appeared independently:

> 'What people call lucky or unlucky omens, and ghosts and spirits, are all produced by the *chhi* of the Great Yang [i.e. the Sun, acting alone]. This solar *chhi* is identical with the *chhi* of Heaven. As Heaven can generate the body of Man, it can also imitate his appearance...When the Yang *chhi* is powerful, but devoid of the Yin, it can merely produce a semblance, but no body. Being nothing but the vital spirit without bones or flesh, it is vague and diffuse, and when it appears it is soon extinguished again.'

In many other places in his book Wang Chhung describes the poisonous and dangerous character of this pure 'form' emanating from the source of all fire and heat, and he considers it the best explanation for all recorded injuries due to supernormal manifestations. This is a naturalist explanation, of which another aspect is seen when he discusses the development of the hen's egg:

'Before a hen's egg is incubated, there is a formless mass within the shell, which, on leaking out, is seen to be of an aqueous nature. But after a good hen has incubated the egg, the body of the chick is formed, and when it has been completed, the chick can pick at the shell and kick [its way out]. Now human death [is a return to] the time of the formless mass. So how could the *chhi* of this formlessness injure anybody?'

In line with his naturalism Wang Chhung saw much scope for chance and strife in Nature, and in order to show how unreasonable it was to insist on relating everything that happens to human beings to their known or alleged moral merits or demerits, he took examples from the non-human world. Thus:

'Mole-crickets and ants creep on the ground. If a man lifts his foot and steps on them, crushed by his weight, they die at once, while those that are untouched continue alive and unhurt. Wild grass is burnt up by fires kindled by the friction of chariot-wheels. People think the tufts of grass which have not been consumed are happy, and call them "lucky grass". Nevertheless, that an insect has not been trodden on, or grass not been reached by a brush fire, is no proof of their excellence.'

He also concluded that the universe, far from being made for the benefit of Man, gave no evidence of detailed design at all:

'If Heaven had produced its creatures on purpose, it ought to have taught them to love each other and not to prey upon and destroy one another. It might be objected that such is the nature of the Five Elements, that when Heaven created all things it imbued them with the *chhi* of the Five Elements, and that these fight together to destroy one another. But then Heaven ought to have filled its creatures with the *chhi* of one element only, and taught them mutual love, not permitting the Five Elements to war against one another and mutually destroy one another.'

As we might expect by now, Wang Chhung has his own thoughts on the creation of the universe, believing that the Earth was formed from a spinning mass of material that later solidified. However, this did not originate with him, for the first inklings of the idea seem to have come from well before the first century A.D. In the *Huai Nan Tzu* (The Book of the Prince of Huai Nan) of 120 B.C., we find:

> 'Before the heavens and the Earth took shape, there was an abyss without form and void; hence the expression "Supreme Light". The Tao began with Emptiness and this Emptiness produced the universe. The universe produced *chhi* [vital, gaseous emanation], and this was like a stream swirling between banks. The pure *chhi*, being tenuous and loosely dispersed, made the heavens; the heavy muddy *chhi*, being condensed and inert, made the Earth.'

The view expressed here was probably derived from the Chinese concepts of condensation and rarefaction mentioned in connection with Taoism (Chapter 8), yet it is probably not the first mention. If a passage in the *Lieh Tzu* is genuine, the view can be taken back to the fourth century B.C.:

> 'We say that there was a great [Principle] of Change, a great Origin, a great Beginning, and a Great Primordial Undifferentiatedness. At the great Change, *chhi* was not yet manifest. At the great Origin *chhi* began to exist. At the great Beginning came the beginning of form and shape. In the great Primordial Undifferentiatedness lay the beginning of matter. When *chhi*, shapes and matter, were still indistinguishably blended together, that state is called chaos. All things were mixed in it, and had not yet been separated from one another... The purer and lighter [elements], tending upwards, made the heavens; the grosser and heavier [elements], tending downwards, made the Earth.'

Wang Chhung's version was very similar:

> 'The *I Ching* commentators say that previous to the differentiation of the original *chhi* there was a chaotic mass. And the Confucian books speak of a wild medley, and of the [two] *chhi* undifferentiated. When it came to separation and differentiation, the pure elements [*chhing chê*] formed Heaven, and the turbid ones [*cho chê*] formed Earth.'

Here, as in previous quotations, the question of exactly *what* rose up and *what* settled down is left vague: the Chinese language made it unnecessary to be precise; *chê* can mean stuff, things, or people, and came naturally to the mind of the writer. This could have been significant if atomism had been acceptable on other grounds. Thus in the twelfth century A.D. we find the great Neo-Confucian Chu Hsi writing about the light and heavy *chhi*, and coming quite close to expressing a particle theory:

> 'The purest [elements] of *chhi* became the sky, the Sun and Man, and the stars, which are permanently revolving and turning round outside. The Earth was in the centre motionless, but not "below".
>
> 'Heaven moves and has moved unceasingly, turning round by day and by night. Earth, that bridge on which we stand, is in its centre. If Heaven were to stop only for an instant, the Earth would collapse. But [in the beginning] the gyration of Heaven was so rapid that there was a great mass of sediment crystallising and coagulating in the middle. This sediment was the sediment of *chhi*, and it is the Earth. Therefore it is said that the purer and lighter parts became the sky, and the grosser and more turbid ones Earth.'

The text certainly trembles on the verge of saying that the sediment was formed by particles made small by mutual friction, though it never quite reaches that point.

We must now turn to consider another of Wang Chhung's attitudes: his opinion of the place of man in the universe. To begin with he made a frontal attack on the Chinese state religion with his uncompromising resistance to anthropocentrism – the doctrine that man is the centre of all things. Again and again he returns to the charge that man lives on the Earth like lice in the folds of a garment, although he does admit that man is the noblest and most intelligent of the naked creatures. But if fleas, he said, wishing to learn of man's opinions, emitted sounds close to his ear, he would not hear them; how absurd then it is to imagine that Heaven and Earth could understand the words of man, or acquaint themselves with his wishes. And once he had gained this position, Wang turned the whole weight of his attack on superstition. Heaven being insubstantial and Earth inert, they can on no account be said to speak or act, so they cannot be affected by anything which man does: they neither hear prayers nor answer questions. This, effectively, swept away the whole basis of divination, whatever the method used.

What now remained of superstition Wang Chhung assailed either by showing how absurd some beliefs were from a statistical point of view, or else how totally unreasonable they were. The thousands of prisoners in the gaols, or all the inhabitants of the city of Li-yang which was flooded during a single night and sank to the bottom of a lake, could not all have chosen unlucky days for their business; nor could the choice of auspicious ones account for all the scholars who attain high official rank.

As for sacrifices to ghosts and spirits, he regarded these as utter nonsense.

> 'The world places confidence in sacrifices, trusting that they procure happiness; and likewise it approves of exorcisms, fancying that these remove evil. The first ceremony performed at exorcising is the setting out of a sacrifice, which we may compare with the entertainment of guests among living men; but after the savoury food has been hospitably set out for the spirits and they have eaten of it, they are chased away with swords and sticks. If the spirits were conscious of such treatment they would surely stand their ground, accept the fight, and refuse to go; and if they were susceptible of indignation, they would cause misfortune. But if they have no consciousness, they cannot possibly effect any evil. Accordingly exorcising is lost labour, and no harm is caused by its omission.
>
> 'Besides it is disputed whether spirits have a material form. If they have, it must be like that of living men. But anything with the form of living men must be capable of feeling indignation, and exorcism would therefore cause harm rather than good. And if they have no material form, driving them away is like [trying to] drive out vapour and clouds, which cannot be done. . .
>
> 'Decaying generations cherish a belief in ghosts. Foolish men seek relief in exorcism. When the Chou dynasty rulers were going to ruin, sacrifice and exorcism were believed in, and peace of mind and spiritual assistance were thus sought. The foolish rulers, whose minds were misled, forgot about the importance of their own behaviour, and the fewer were their good actions the more unstable their thrones became. The conclusion is that man has his happiness in his own hands, and that the spirits have nothing to do with it. It depends on his virtues, and not on sacrifices.'

Wang Chhung's fervour rises almost to a prophetic level reminiscent of Isaiah, and it was, perhaps, rather characteristic of Chinese civilisation that if we have to look anywhere for a parellel of the moral force of the Hebrew prophets, it would be found among the most atheist and agnostic of Confucian rationalists.

Another set of ideas that Wang Chhung attacked was the Taoist belief in the attainment of material immortality by the aid of physical and mental techniques. His arguments are interesting. First he compares the

Taoist aim and its longevity practices with biological metamorphosis. For example, he says that while quails and crabs 'metamorphose', this does not prevent them from being eaten, and the life span of insects that metamorphose is very short and compares unfavourably with animals that do not metamorphose. Wang's zoology may have been rather primitive, for his 'metamorphosis' included all kinds of changes, some real but some quite untrue though generally believed in his day, yet his arguments are no less effective for that. Secondly, he pointed out that while peace of mind might be considered by the Taoists to be helpful to longevity, what of plants and herbs? Though they are quite dispassionate, peace of mind brings them no longevity, for they often live for no more than a year. Again, living organisms can be harmed by too much ventilation, so what is the purpose of respiratory exercises? And if one looks at a river and sees how turbulent it becomes as it flows through the land, why try to increase the flow of blood by gymnastics? Clearly, the blood-stream will be purest if not interfered with. Wang finally strengthened his case still further by technical arguments against alchemy.

All in all, Wang Chhung, as might be expected of a serious Confucian, was in favour of Taoist naturalism but against their experimental techniques, although perhaps he changed his mind a little as he grew older. Nevertheless, he always remained a staunch enemy of any belief in spiritual beings, and fought with fervour against the vast body of legends about supernatural births, intercourse with dragons, and so on, which most of his contemporaries still seriously believed.

Wang Chhung and the Phenomenalists

Wang Chhung agreed however with the Taoist view of the existence of a better, more primitive era, and he used this as a base from which to launch an attack on the Phenomenalists.

'Originally there were no calamities or omens, or if there were, they were not considered as reprimands [from Heaven]. Why? Because at that time people were simple and unsophisticated, and did not restrain or reproach one another. Later ages have gradually declined – superiors and inferiors contradict one another, and calamities and omens constantly occur. Hence the hypothesis of reprimands [from Heaven] has been invented. Yet the Heaven today is the same Heaven as of old – it is not that Heaven anciently was kind, and now is harsh. The hypothesis of Heavenly Reprimands has been put forward in modern times, as a surmise made by men from their own [subjective] feelings.'

This was a frontal assault on a very deep-rooted set of beliefs and a powerful group of people, namely those Han Confucians who had developed the ideas of the School of Naturalists to produce a system in which any ethical irregularity had cosmic effects. According to them, if the emperor and his ministers did not practise their rites and ceremonies properly then excessive gales would follow and trees would not grow properly; if the emperor's speech was not in accord with reason, metals would cease to be malleable – and so on through a whole hierarchy of effects. Moreover, this applied not only to the emperor, but to the whole of his bureaucracy, faults of local officials causing a host of local natural irregularities.

Phenomenalism was essentially a kind of inverted astrology, and the flood of literature to which it gave rise, devoted to the detection and interpretation of the meaning of all abnormal and catastrophic phenomena in the skies or on the Earth, left behind it a mass of débris in every one of the Chinese dynastic histories. Indeed, in Wang Chhung's time even the classics were being searched for material to fit in with the theory, and when this proved insufficient, as it did, new texts were invented. In due course these new texts became considered authoritative, and were elevated to such a degree of importance that, in the first and second centuries A.D., many important state matters were decided solely by reference to them. To all this Wang led a vehement opposition, bringing forward every argument he could muster to counter, for instance, such ideas as that seasonally excessive heat or cold depend on the ruler's joy or anger, or that plagues of tigers and grain-eating insects are a result of the wickedness of secretaries and minor officials. He emphatically denied that natural calamities and unlucky events were the result of Heaven's anger, or hard winters the outcome of cruelties and oppressions. All the events supposed to be due to administrative errors were, he said, due purely to chance:

> 'The setting in of torrid or frigid weather does not depend on any governmental actions, but heat and cold may chance to be coincident with rewards and punishments, and it is for this reason that the Phenomenalists [falsely] describe them as having such a connection.'

Here we meet again with the belief in pre-established harmony. The Phenomenalists thought they had detected invariable manifestations of it; Wang Chhung was convinced they had done nothing of the sort.

Wang's denial of phenomenalism involved him in another serious question, that of 'action at a distance'. If Man's actions and Heaven's response did not exist, and especially if Man was not the centre of all things, what of the action of one thing on another in the context of the

harmony of the organism of the universe? A purely material sequence
of cause and effect was certainly ruled out, and Wang Chhung was
obliged to admit some form of action at a distance; what he did, though,
was to limit its effect. A dragon can cause rain, but only within a
distance of 100 *li* (some 50 km); there may be telepathy but not beyond
a certain distance; and he did admit that the influences emanating from
the stars constituted an important endowment of individual human
beings. Certainly Nature could affect Man, but it was, he claimed, asking
too much to suppose that the puny doings of Man could affect Nature.
Unfortunately, though, Wang Chhung's protests were not very effective:
phenomenalism continued to flourish, and when it did at last peter out,
it was primarily for reasons other than his criticisms. It was only after
a time, when one rebellion after another claimed justification from the
new texts, that bureaucracy began at last to frown on them.

Wang Chhung and human destiny

Wang Chhung was a stern critic of the errors of his contemporaries,
but his contribution to Chinese culture also had a positive side. Particularly was this so when it came to human affairs and the question of
fate. His conception of destiny was not of an inexorable decree of Heaven
laid down for each individual, but of three separate factors: first, a
spiritual essence with which each individual human constitution was
endowed; secondly specific influences emanating from the stars, and,
thirdly, the effects of chance.

As far as the spiritual essence of each individual was concerned, Wang
Chhung meant not only mental endowment but also something
inherently physical. 'The fate of individuals', he wrote, 'is inherent in
their bodies, just as with birds the distinction between cocks and
hens exists already within the eggshell.' It was an insistence, and an
interesting one, on genetic inheritance as against environmental influences, and it led him to accept the teachings of physiognomy. The second
factor, the astrological, is also important. When Wang Chhung was
propounding his ideas astrology was beginning to spread out from the
courts to the mass of people, and it seems that it was his strong scientific
naturalism that pushed him into his belief in a new dimension to
astrology, the personal horoscope. It was a means of escaping from the
arbitrary effects supposedly caused by local gods and spirits, or by other
supernatural agencies. Certainly Wang would have no truck with some
traditional accounts of the behaviour of celestial bodies that he found
absurd, accounts like that which stated that on the occasion of a certain
battle the Sun moved backwards in the sky through three mansions
(almost 40 degrees), or another which claimed that Mars moved forward
by a similar amount after a feudal lord had uttered some particularly

excellent maxims. All the same, whatever the reasons, it may turn out a paradox in the history of science in China, that astrology for the individual was founded by the greatest of all the sceptics of pseudo-science.

The last of the three factors was chance, which Wang Chhung tried to analyse further. He distinguished between the effects of time and contingencies such as general calamities where many people perish at once; he also differentiated these from luck, as in the arrival of an amnesty after a man has been imprisoned, and from incidents such as chance encounters with men who have high official posts in their gift. Wang never succeeded in establishing proper definitions, but his views were certainly different from those of his contemporaries, who recognised only three divisions, not four – natural fate, an evil will aiding an evil fate and vice versa, and an evil will acting against a good fate or good fate against an evil will. He also believed in a strict pre-determined pattern in which the will of the individual could do little or nothing.

'In eulogy we say' (to use a stock phrase of the dynastic histories) that Wang Chhung was one of the greatest figures of his age from the point of view of the history of scientific thought. As he himself wrote:

> 'One sentence is enough to sum up my book – it hates
> falsehood. Right is made to appear wrong, and falsehood is
> regarded as truth. How can I remain silent? When I read
> current books of this kind, when I see the truth overshadowed
> by falsehood, my heart beats violently and the pen trembles in
> my hand. How can I be silent? When I criticise them, I study
> them, check them against facts, and show up their falsehood
> by appealing to evidence.'

Unfortunately, the main value of Wang Chhung's work was negative and destructive. If only he could have devised some hypothesis more fruitful for science and technology than the Yin and the Yang, and the Five Elements, his services to Chinese thought would have been greater still.

The sceptical tradition in later centuries

The sceptical tradition that Wang Chhung emphasised runs on throughout Chinese history. Indeed, it stands out as one of the great achievements of Chinese culture when one compares it with the miasma of religious, theological and magical writings dominant in some other civilisations. Because of it, Confucians adopted the attitude of ridiculing a belief in spirits, and when the growing power of Buddhism brought great reinforcements to the side of superstition, there were always

Confucians to oppose it. When, for instance, in A.D. 484, a great debate
was held before the Prince of Ching-Ling, the sceptical philosopher Fan
Chen attacked the Buddhist doctrine of *karma* (the explanation of good
and evil in this life in terms of good and bad actions in previous lives),
keeping Wang Chhung's well-known attacks on the belief in immortality,
which reincarnation implies, well to the fore. Fan Chen then made his
celebrated statement that 'The spirit is to the body as the sharpness is
to the knife. We have never heard that after the knife has been destroyed
the sharpness can persist.' Later, his views were embodied in an essay
which so alarmed the Buddhists that more than 70 refutations of it were
written.

Other records show us the tradition continuing in action. In 632 Lü
Tshai, a figure reminiscent in many ways of Wang Chhung, was ordered
to edit the existing books of divination and of the Yin–Yang and
Five-Element theories, and to add a sceptical preface to each. Stories
directed against superstition of all kinds now became more frequent and,
some seven centuries later, the astronomer and astrologer Liu Chi was
still to be found seeking to explain natural phenomena by purely natural
causes. One of his contemporaries, Hsieh Ying-Fang, went so far as to
make a collection of anti-superstitious material in A.D. 1348 in his *Pien
Huo Pien* (Disputations on Doubtful Matters). In Early Ming times
other exponents of the sceptical tradition are to be found, while the Late
Ming saw the distinction between Taoist empirical pseudo-science and
the sceptical Confucian tradition well and clearly defined. But this was
the end of the road, for it was then that the 'new or experimental
philosophy' of the West reached Peking.

Chinese humanistic studies as the crowning achievement of the sceptical tradition

As a kind of appendix to what we have so far said about the
sceptical tradition, we ought to glance at some other developments to
which it gave rise – humanistic studies, textual criticism and archaeology. These were all fields in which the followers of this tradition could
find full outlet with fruitful results, since the evidence was to hand, and
no failure to treat scientific theories in a mathematical way entered into
the matter.

China became the first home of the humanistic sciences, for although
there was some critical study and dating of ancient texts in the West
at the great library in Alexandria in the second century B.C., this was
an isolated case, and Western studies along these lines did not really
get going until the late seventeenth century. In China linguistic studies
go back to Han times when scholars applied some of Wang Chhung's

scepticism to the examination of ancient texts, a study encouraged by the state in view of the great store that was set by the classics. But the real flowering of Chinese critical humanism came during the tenth to the thirteenth centuries, at the same time as a peak of Chinese activity in all branches of science and technology, and the rise of Neo-Confucianism, that great philosophical achievement of a scientific view of the world. All this was at a time when Europe could show nothing even remotely comparable.

Perhaps one of the triggers that set the movement going was the dissatisfaction there was with the texts invented to support phenomenalism (see p. 210). Once these texts had been examined and their lack of authenticity established, scholars turned their critical attention to the really ancient texts and the traditional commentaries on them. Compendia of the best opinions on the ancient texts were later compiled, one of the most famous, the *Chün-Chai Tu Shu Chih* of Chhao Kung-Wu, which appeared in 1175, being so good that it was heavily relied on by the compilers of the Imperial Manuscript Library catalogue six centuries later. After the fall of the Mongols and throughout the Ming, the movement was largely in abeyance, perhaps because a rise of patriotism militated against criticism of what were considered semi-sacred national texts, but at the end of the sixteenth century, after a lapse of two hundred years, critical historical analysis began once more, producing results that were in no way inferior to the famous school of English criticism in the West a century later.

This critical tradition was accompanied by an interest in archaeology, of which the earliest efforts go back a long time, so that by the Sung (tenth century A.D.) the subject had risen to a thoroughly scientific level. In the eleventh century studies of inscriptions (epigraphy) were also well under way, and there appeared what was probably the first book in any language on the subject. The twelfth century was even more productive: in the opening year the Emperor Hui Tsung came to the throne and straight away embarked on the formation of an archaeological museum, of which a detailed catalogue was issued a dozen years later. Then, in 1134, a book appeared on the origin of family names, and in 1149, Hui Tsung himself published his *Chhüan Chih* (Treatise on Coinage), without doubt the first book in any language on numismatics. Other pioneering efforts followed: in 1307 a study of jade appeared, and in spite of a Ming interregnum, archaeology arose again as the seventeenth century opened, to become one of the glories of Chinese scholarship.

Thus we see that the sceptical tradition of China was not merely empty and theoretical, not just conventional and destructive. The Confucians did not share an interest in the non-human world of Nature with the

Taoists, but within the domain of human life and thought lay a field open for the application of scientific method without experimentation, and this they made their own. What came out of it was not that terrifying strength that has transformed the natural world, but rather a vast edifice of knowledge of the past history of a people, an edifice to which its European counterpart has only been even comparable for two centuries past.

12

Chin and Thang Taoists and Sung Neo-Confucians

Taoist thought in the Wei and Chin periods

It seems to have been the primitive scientific teaching of the Taoists that was the first aspect of Taoism to die out for want of appreciation. Other Taoist ideas were discussed and commented on, but their proto-science went by default. By the time of the Wei and Chin dynasties (third and fourth centuries A.D.), the Confucians were busy encouraging what one might call a thorough revisionism of Taoist philosophy in favour of religious and mystical interpretations, probably because anything else would have threatened their own supremacy and the stability of their bureaucratic order. Both scientific observation and the extremely democratic ideals of the Taoists were seen as something shocking, and a gross distortion of the Taoist philosophers consequently took place. A 'Mystical School' (*Hsüan Hsüeh*) developed, led by the new commentators of Taoist books, all of whom were men who considered the Confucian sages far more worthy than any Taoist, although one or two of them did practise arts like alchemy. The Tao was now interpreted as 'non-being', and Confucius was praised because he had not tried to speak about the unspeakable. The words of Chuang Chou were also praised but considered useless for application to human society, while the peace gained by the Taoists in contemplating Nature was totally misunderstood. It was all an attempt to turn the original theories of Taoism into a philosophy acceptable in a milieu where Confucianism was dominant.

As a result of all this re-assessment, mixed Taoist-Confucian schools grew up, and in these 'Philosophic Wit' or 'Pure Conversation' groups, epigrams were prized and mundane matters avoided. Nevertheless, the schools attracted the best intellects of the time, and it was not long before some among them began to react strongly against Establishment opinions, stressing above all the social radicalism of the Taoist outlook. One of these whose name has come down to us was the poet Hsi Khang, who scandalised the Confucians by his skill as a metal-worker. However, his refusal, and the refusal of others like him, to accept the restraints of conventional morality and social institutions went far beyond what was allowable, almost reaching the fringes of an attitude for which moral institutions no longer mattered, and only one's own faith was of any significance. Or so it is reported, but we have few of their own writings from which to form our own opinion, and must remember that what we do read was written by their enemies.

During this period the revisionists did not have it all their own way. Not only were there rebels like Hsi Khang, but even some of the early commentators themselves still appreciated a good deal of the political position of the ancient Taoists. Then, in the late third and early fourth centuries, there arose the singular figure of Pao Ching-Yen, the most radical thinker of all the mediaeval Chinese centuries. We know nothing of his life, for he apears only in conversation with the alchemist Ko Hung in a long chapter of the latter's *Pao Phu Tzu* (Book of the Preservation-of-Solidarity Master) published early in the fourth century. In Pao Ching-Yen the old Taoist aversion to feudalism, now slightly modified to face the feudal bureaucracy of the time, seems to have lost nothing of its force.

> 'The Confucians say that Heaven created the people, and
> planted lords over them. But why should illustrious Heaven
> be brought into the matter, and why should it have given such
> precise instructions? The strong overcame the weak and
> brought them into subjection, the clever outwitted the simple,
> and made them serve them – this was the origin of lords and
> officials, and the beginning of mastery over the simple
> people... Heaven had nothing whatever to do with it...
> 'In ancient times there were no lords and officials. Man
> [spontaneously] dug wells for water, and ploughed fields for
> food. Man in the morning went forth to his labour [without
> being ordered to do so] and rested in the evening. People were
> free and uninhibited and at peace; they did not compete with
> one another, and knew neither shame nor honours...'

and the text goes on to describe the decrease of simplicity and honesty and the growth of artificiality, in traditional Taoist style. Pao Ching-Yen also contrasts the ideal Taoist state with the actual state of his own day: the desires of the lords are insatiable; they monopolise women in the inner apartments, and throw away in useless extravagances money derived from the bitter labours of the people. In the stores of the lords and officials there is abundant food and clothing, but the people go hungry and cold. So long as these inequalities persist, all laws and ordinances, however just, will be valueless. Pao Ching-Yen, whoever he was, had obviously a clear insight into social oppression and the origins of social strife, fully sounding once again the political note of the early Taoists. But he was almost the last to do so.

In contrast to the declining fate of its political and philosophical doctrines, the experimental traditions of ancient Taoism not only continued but flourished. Throughout the period of the Three Kingdoms and the Chin (about A.D. 270) to the beginning of the Sui dynasty (A.D. 580) alchemy was cultivated to a hitherto undreamed-of extent. It was during this time that Ko Hung, the greatest alchemical writer in Chinese history, lived and worked, but it is a period about which we are poorly informed, because after the invention of paper during the Han, a great many books were written on what proved to be non-durable material. We are fortunate, though, in having Ko Hung's *Pao Phu Tzu*, for the earlier chapters contain some scientific thinking at a high level. Here we see a mind groping its way towards a mastery of the complexities of Nature, a mind more impressed by diversity of phenomena than with generalisations to unify them. The book was also written in a beautiful prose style, and this commended it to scholars whose Confucianism would not otherwise have let them pay attention to it. To glimpse a sample, let us now break into the text where an argument is going on about the possibility of achieving immortality:

> 'Someone said to Ko Hung: "Even [Lu] Pan [legendary mechanic of the state of Lu] and [Mo] Ti could not make sharp needles out of shards and stones. Even Ou Yeh [legendary metallurgist] could not weld a fine blade out of lead and tin. The very gods and spirits cannot make possible what is really impossible; Heaven and Earth themselves cannot do what cannot be done. How is it possible for us human beings to find a method which will give constant youth to those who must grow old, or to revive those who must die? And yet you say that [by the power of alchemy] you can cause a cicada to live for a year, and an ephemeral mushroom to survive for many months. Don't you think you must be wrong?..."
>
> 'Pao Phu Tzu answered "Things which in the end appear different may have the same root and origin. Things cannot all be spoken of in one way. Things which have a beginning generally also have an end, but this is not a universally applicable principle. Thus it may be said that everything grows in summer, but shepherd's purse [*Capsella Bursa-pastoris*]

and wheat fade then. It may be said that everything withers in winter, but bamboos and the arbor-vitae bush [*Thuja orientalis*] flourish then. It may be said that everything comes to an end as it begins, but Heaven and Earth have no end. It is generally said that life is followed by death, but tortoises and cranes live almost for ever. In summer the weather is supposed to be hot, but we often have cool days then; in winter the weather is supposed to be cold, but mild days occur. A hundred rivers flow to the east, but one large river flows to the north. The Earth by nature is quiet, but sometimes it trembles and crumbles. Water by nature is cool, but there are hot springs in Wên-Ku. Fire is by nature hot, but there is a cool flame upon Hsiao-Chhiu mountain. Heavy things ought to sink in water, but in the South Seas there are floating hills of stone. Light things ought to float, but in Tsang Kho there is a stream in which even a feather sinks. No single generalisation can cover such multitudes of things, as these examples show. . . Thus it is not to be wondered at that the *hsien* does not die like other human beings."

'Someone else said: "It may be admitted that the *hsien* differs very much from ordinary men, but just as the pine tree compared with other plants is endowed with an extremely long life, so may not the longevity of the *hsien* exemplified in Lao Tzu and Phêng Tsu be after all a [special] endowment from Nature? One cannot believe that anyone could *learn* to acquire longevity such as theirs."

'Ko Hung replied: "Of course the pine belongs to a kind different from other trees. But Lao Tzu and Phêng Tsu were human beings like ourselves. Since they could live so long, we can also."'

Here, apart from references which explain themselves, Ko Hung may have been talking about natural gas flames, light petroleum seepages and floating islands. Admittedly, there is much in the *Pao Phu Tzu* book that is wild, fanciful and superstitious, just as there is in Paracelsus twelve centuries later, but the arguments are logical and the vision broad. It was rather better too than anything written in the West at this time by the Greek proto-chemists. One more passage will illustrate the mixture of strange beliefs and true facts which was so characteristic of Ko Hung and of the other Taoist alchemists:

'As for the art of Change, there is nothing it cannot accomplish. The body of man can naturally be seen, but there are means to make it invisible; ghosts and spirits are naturally invisible yet there are means whereby they can be caused to appear. These things have been repeatedly done.

'Water and fire, which are in the heavens, may be obtained by the burning-mirror and the dew-mirror. Lead which is white, can be turned into a red substance. This red substance can again be whitened to lead. Clouds, rain, frost, and snow, which are all the *chhi* of Heaven and Earth, can be duplicated exactly and without any difference, by chemical substances.

'Creatures which fly and run, and creatures which crawl, all derive a fixed form from the Foundation of Change. Yet suddenly they may change the old body and become totally different things. Of these changes there are so many thousands and tens of thousands that one could never come to the end of describing them.

'Man is the noblest of all creatures, yet men or women may be transformed into cranes, stones, tigers, monkeys, sand or turtles. Similarly the transformation of high mountains into abysses, and the making of peaks out of deep valleys, are examples of change in huge things. Change is inherent in the nature of Heaven and Earth. Why then should we think that gold and silver cannot be made from other things?...

'Narrow-minded and ignorant people take the profound as if it were uncouth, and relegate the marvellous to the realm of fiction. To these people anything that was not spoken of by the Duke of Chou and Confucius, and not mentioned in the classics, is untrue. What narrow-mindedness and ignorance!'

Ko Hung certainly lived at a time that was sympathetic to his kind of outlook, a time that even after his death still valued alchemy so highly that between A.D. 389 and 404 the Northern Wei emperor could establish at the capital both a professorship of Taoism and a Taoist laboratory for the making of medicinal preparations. Indeed the Western Mountains were allocated to supply firewood for the alchemical furnaces, and those guilty of capital offences were made to test the elixirs in person; all this in spite of efforts by the Imperial Physician to have the laboratory closed down.

During these centuries, just as the Confucians reinterpreted the books of Taoist philosophy, so the Taoists appropriated the *I Ching* (Book of Changes), elaborating it in an attempt to devise a general scientific theory. This movement seems to have begun in the Later Han, since A.D. 142 is the date of its pioneering work, Wei Po-Yang's alchemical treatise, the *Chou I Tshan Thung Chhi* (Book of the Kinship of the Three; of the Accordance of the *Kua* of the Book of Changes with the Phenomena of Composite Things) more usually known as the *Tshan Thung Chhi*. In this, as we saw in Chapter 10, there was a complex system of associations between the eight trigrams and the cycle of 'stems', with the aim of symbolising the various stages of the movements of the sun and moon, and hence the waxings and wanings of the Yin and Yang influences in the world. The alchemists considered the book and the system it contained to be of great importance in determining their choice of the precise times for performing their heating and other operations.

Taoist thought in the Thang and Sung periods:
Chhen Thuan and Than Chhiao

The kind of thinking contained in the *Tshan Thung Chhi* appears to have reached its climax with Chhen Thuan, a somewhat shadowy figure who rose to great prominence during the short-lived dynasties between the Thang and the Sung. He took his degree in the Later Thang in 932, was first called to court under the Later Chou in 954, and between 976 and 984 was treated with great honour by the second Sung emperor. At the end of this period, when he must have been quite elderly, he pleaded ignorance and retired as much as he could to a solitary way of life, dying five years later. There is much about him and his work in the *I Thu Ming Pien* (Clarifications of the Diagrams in the Book of Changes), a book written by Hu Wei in 1706. In spite of its late date this is a work of great importance, because it gives a careful and critical history of the alchemical philosophy connected with the *I Ching*, and destroys the traditional view that it went back to remote antiquity: indeed it is still essential for studying the development of Chinese thought.

According to Hu Wei, it was Chhen Thuan who started the specific arrangements of the trigrams in the *I Ching*, although it is clear too that he owed much to Wei Po-Yang and his *Tshan Thung Chhi*. But from our point of view at this stage, since we are not at the moment dealing with alchemy, what is significant is that we see how the *I Ching* hexagrams were being thought of as embedded deep in the fabric of Nature. This was characteristic of mediaeval Chinese scientific thinking, but as previously mentioned, it was a factor tending to inhibit truly scientific interpretations of Nature. Nevertheless it was not without its significance, for it prepared the way for a new Confucian synthesis during the Sung, and this not merely because the Taoist liking for diagrams stimulated the Sung thinkers to make their own, but also because it emphasised the thoroughgoing naturalness of the world. For these mediaeval Taoists believed that there was no force in Nature which Man could not control, provided he knew the right techniques. If there were a spiritual element, it was not so overwhelming that Man could only bow down before it, but rather a collection of spirits within the natural order, which Man could command to serve him.

During the Thang dynasty there was also a second flowering of true Taoist philosophy. Between the sixth and the tenth centuries there were many books which, using the new Taoist experimental researches as a background, revived and expanded many of the old doctrines. Typical is the *Kuan Yin Tzu* book, also known as the *Wên Shih Chen Ching* (True Classic of the World), written by an unknown Taoist late in the Thang or just after the dynasty's end, and edited with a commentary by Chhen Hsien-Wei, who also called himself Pao I Tzu. Here we read:

> 'Man's might can conquer the changes of Nature, make thunder in winter and ice in summer, make the dead walk and the dry wood blossom, confine a spirit in a bean, and catch a [big] fish in a cupful of water, open doors in paintings and make images speak. It is pure *chhi* which changes the myriad things; where it agglomerates it causes life; where it disperses it causes death. What has never been aggregated or dispersed has never been alive or dead. The guests [living things, or phenomena] come and go, but their material basis remains unchanged.'

This last is a very scientific statement, especially in its almost classic definition of a materialism where all changes are but apparent, due to the combinations and recombinations of elements which themselves never change:

> 'The changes occurring in the myriad things are all due to the *chhi*, but whether they are hidden or whether they can be seen, the *chhi* remains a unity. The sage knows that the *chhi* itself is one, and never changes.'

One is almost tempted to see in such a statement a premonition of the nineteenth-century 'first law of thermodynamics' (the conservation of energy).

Throughout these Thang writers there is an appreciation of the subtle reactions of Man and Nature, each affecting the other. This still sometimes took an ancient phenomenalist form, but often it reached deeper levels. Then they discussed questions of potentiality and actuality, the *Kuan Yin Tzu* book saying:

'The *chhi* involves a time factor. That which is not *chhi* knows no day nor night. Form has a spatial factor. That which has no form has no south nor north. What is "not-*chhi*"? It is that which produces *chhi*. For instance, if a fan is agitated, a wind is produced and the *chhi* becomes palpable as the wind. What is "not-form"? It is that which produces form. For instance, if wood is bored [to make fire], fire is produced, and form becomes visible as fire.'

This is of considerable interest, because it shows the Thang Taoists groping after something more fundamental in the universe than *chhi* (matter) and *hsing* (form): their forefathers would have been content with the term Tao, but now something more precise was needed. It came in due course, as we shall see, with the Neo-Confucians, but even before this we find the Taoists appreciating the fact that actualisation, coming-into-being, growth, is much more prolonged and apparently a more difficult than decay. As Master Kuan Yin put it:

'The construction of things is difficult; the destruction of things by the Tao is easy. Of all things under Heaven there is none that does not reach its completion with difficulty, and none that is not easily destroyed.'

– a passage that reminds one of William Harvey's seventeenth-century comment on embryology:

'For more, and abler, operations, are required to the Fabrick and erection of Living creatures, than to their dissolution, and the plucking of them down; for those things that easily and nimbly perish, are slow and difficult in their rise and complement.'

Some appreciation of the methods of induction and deduction may also be found in the Taoist books. To quote the *Kuan Yin Tzu* again:

'Ordinary people are bewildered by names; they see the things but not the Tao of each thing. The [Confucian] worthy analyzes principles; he sees the Tao but not the individual things. But the true [Taoist] sage unites himself with Heaven; he sees neither the Taos nor the things, for one Tao includes all the separate Taos. If you do not apply it to individual things you reach the Tao of all things – if you apply it to the individual Tao's then you understand the things.'

Here Master Kuan Yin seems to be saying that while ordinary people are not interested in generalisations, the Confucian is interested in assuming them to begin with, but the Taoist, in seeking neither generalisations nor individual phenomena, sees in fact both, looking sometimes at the particular and sometimes at the general. Again:

'Those who understand the Tao [to which I refer] will know the Way of Heaven, understand the divine power of Nature, comprehend the destinies of things, and penetrate Nature's mysteries; all from observation of phenomena. Having this Tao you can also unite all different results, and forget all the different names [of the diverse phenomena].'

Nevertheless, the old distrust of logic and abstract reasoning still dogged the Taoist's footsteps:

'The wisest man of all knows that human knowledge cannot grasp the things of Nature, therefore he looks as if he were foolish. The best dialectician of all knows that argument will not succeed in describing the things of Nature, therefore he seems to stammer. The bravest man of all knows that courage cannot overcome the things of Nature, therefore he seems to be afraid.'

And the perennial theme of the sage being without partiality was still in evidence:

'The sages learnt social order from bees, textiles and nets from spiders, ceremonies from praying rats, and war from fighting ants. Thus the sages were taught by the myriad things, and in their turn taught the worthies, who taught the people. But only the sages could understand the things [in the first place]; they could unify themselves with natural principles, because they had no prejudices and preconceived opinions.'

Nor had the old scientific view of the universe been lost, even though little progress had been made in scientific theory. The eighth-century *Yin Fu Ching* (Harmony of the Seen and Unseen) says:

'The spontaneous Tao [operates in] stillness, and so it was that Heaven, Earth and the myriad things were produced. The Tao of Heaven and Earth [operates like the process of] steeping [i.e. when chemical changes are brought about gently, gradually and insensibly].

'[Thus it is that] the Yin and the Yang alternately conquer each other, and displace each other, and change and transformation proceed accordingly.

'Therefore the sages, knowing that the spontaneous Tao cannot be resisted, follow after it [observing it], and use its regularities. Statutes and calendrical tables drawn up by men cannot embody [the fullness of] the insensibly acting Tao, yet there is a wonderful machinery by means of which all the heavenly bodies are produced, the eight trigram symbols, and the sexagenary cycle. This is a spiritual machinery indeed, a ghostly treasury. All these things, together with the arts of the Yin and the Yang in their mutual conquests [to him who understands the Tao] come forward into bright visibility.'

The co-operative politics of the old Taoist philosophers were not forgotten either, but from the scientific point of view it is, perhaps, more interesting to note the mixture of magic, experimentation, bodily culture and the invulnerability complex, as also the persistence of the old theme that techniques should be used for understanding Nature rather than for benefiting human society. Thus the *Kuan Yin Tzu* book:

'There are in the world many magical arts; some prefer those that are mysterious, some those which are understandable, some the powerful ones, some the weak ones. If you grasp [apply] them you may be able to manage affairs; but you must let them go in order to attain the Tao.

'The Tao originates in Non-Being... Things originate in Being, but the Tao controls their hundred actions. If you attain the height of the Tao you can benefit humanity; if you attain the loneliness of the Tao you can establish your personality. If you realise that the Tao is not in time, you can take one day as a hundred years and conversely. If you know that it is not in space, you can reckon one *li* as a hundred *li* and conversely. If you know that the Tao, which has no *chhi*, controls the things which have *chhi*, you can summon the wind and the rain. If you know that the Tao, which is formless, can change the things that have form, you can change the bodies of birds and animals. If you can attain the purity of the Tao, you can never be implicated in things; your body will feel light, and you will be able to ride on the phoenix and the crane. If you can attain the homogeneity of the Tao, nothing will be able to attack you; your body will be dark, and you will be able to caress crocodiles and whales...

'If you know that the *chhi* emanates from the mind you will be able to attain spiritual respiration, and you will succeed with alchemical transmutations of the stove...

'If you unify yourself with all things, you can go unharmed through water and fire.

'Only those who have the Tao can perform these actions – and, better still, not perform them, though able to perform them!'

How applicable are these final words to our present-day state of high technology, so dangerous as well as so promising for mankind.

Of all the books of the period, the most original from the point of view of the philosophy of science is probably the *Hua Shu* (Book of the Transformations in Nature), supposedly written by Than Chhiao, in the tenth century. Here there is a special kind of subjective realism – the external world is real but our knowledge of it is deeply affected by the way it is perceived, so that we cannot seize its full reality. For example, to the owl the night is bright and the day dark; for the hen the converse is true, as for ourselves. Which of the two, Than Chhiao then asks in good Taoist style, is to be considered 'normal' and which 'abnormal'? The answer, he says, is that one cannot assume that daytime is bright and fit for sense-perceptions, while the night is not – it depends on the nature of the sense organs. The inference here is that the colours we see and the sounds we hear are not really present, but are constructs of our own organs of sense: almost an anticipation of the distinction between primary (real) and secondary (subjective) qualities to be reached by the European philosopher John Locke some eight centuries later. Then Chhiao next refers to optical illusions and human attention, or inattention, to what is seen. A man, he says, may shoot at a striped stone under the impression that it is a tiger, or at a ripple in the water in the belief that it is a crocodile. Moreover, if the animals really are there, his attention will be so concentrated on them that he will not notice the stones or the water beside them. This leads him to infer that nothing is real in the sense that it demands to be perceived, and we only pick out certain elements to form our world-picture. This can be extended to life and death themselves, he says. Only the Tao (the substratum of all sense-impressions of all beings) is truly real. Everything else is relative and our sense-organs can therefore never give us an absolute picture of the external world.

As to cause and effect in Nature, Than Chhiao was clear that determining factors had to be looked for, factors that he thought might be quite inconspicuous:

'The control of a ship carrying ten thousand bushels of
cargo is assured by means of a piece of wood no more than
a couple of metres in length [the rudder]. The action of a
crossbow-catapult of one thousand *chün* [a pull of more
than 20 tonnes in weight] depends upon an apparatus [the
trigger] which is no more than a couple of centimetres in
length. With one eye one can see the vast expanse of the
heavens; and millions of people can be governed by one
emperor. . . If you can realise the connectedness of Heaven
and Earth; if you can understand the "field of force" of the
Yin and the Yang, if you can know the hidden storehouse
of seminal essence and spirit; then you can overcome [i.e.
change] the numbers [written in the book] of destiny, you
can prolong your life, and you can turn all things upside
down [i.e. control Nature].'

The origins of Neo-Confucianism and the Sung Neo-Confucians

Thus under the Thang dynasty Taoism flourished, but it must
not be imagined that the Taoists had it all their own way. They had
to contend not only with the Buddhists, but also with a number of
Confucians who were re-thinking matters, men whose views were to play
an important part in founding the new philosophical school of Neo-
Confucianism that was to become so powerful an intellectual force in
Sung times. Of these the most significant philosophically was Li Ao
(d. A.D. 844), who in his *Fu Hsing Shu* (Essay on Returning to the
Nature) used a number of technical terms that were to become important
in Neo-Confucian thinking. The book also contains one remark that is
rather revealing about the origins of Neo-Confucianism:

'Although writings dealing with [human] Nature and with
Destiny are still preserved, none of the scholars understand
them, and therefore they all plunge into Taoism or Buddhism.
Ignorant people say that the followers of the Master
[Confucius] are incapable of investigating the teachings on the
Nature and Destiny, and everybody believes them.'

This strongly suggests that during Thang times the Confucians began
to feel acutely the lack of any cosmology to offset that of the Taoists,
or of a metaphysics to compete with that of Buddhism. Sung Neo-
Confucianism, a system made by borrowing various elements from the
other two schools, was evolved to satisfy this need, and in doing so
rescued the ethical teaching of the classics from threatened oblivion by
bringing it into close contact with a reasoned theory of the universe.

The value of Taoism lay in its naturalism; its defect was its failure to take much interest in human society. Recognising that ethical considerations were irrelevant to scientific observations or to scientific thought, it had offered no explanations of how the highest human values manifested in society could be related to the non-human world. As Hsün Chhing had said, 'They see Nature, and fail to see Man.' On the other hand the Buddhists' metaphysical idealism, concerned, as its name implies, with the underlying nature of all creation, was one stage worse, for it was interested neither in Nature nor in human society. The Buddhist saw both as elements of a giant conjuring trick from which all beings must be helped to escape; but to look on the world as a giant phantasmagoria does not induce scientific study nor encourage public justice. Yet there was no sense in returning to antique Confucianism, for its total lack of cosmology and philosophy meant that it could no loner satisfy a maturer age. There was, in fact, only one way out, and this was the way taken by the Neo-Confucians: to use a prodigious effort of philosophical insight and imagination to set the highest ethical ideals of man in their proper place against the background of non-human Nature, or rather within the vast framework of Nature as a whole. On such a view the nature of the universe is in one sense moral, not because a moral personal deity exists somewhere outside space and time directing it all, but because the universe has the property of bringing forth moral values and moral behaviour when the appropriate level of organisation has been reached. Modern evolutionary philosophers look on this idea of development as one long continuous process; the Neo-Confucians thought there were many successive developments, each occurring after a world conflagration, but the basic conception seems to have been much the same.

Western scholars studying Neo-Confucianism have been too often perplexed by their own failure to appreciate this. The Jesuits who went to Peking were outraged because the Neo-Confucians denied a personal God, while Protestant theologians later claimed to detect in Neo-Confucianism a kind of pantheism, equating God with Nature. Indeed one even suggested that the materialism of the Neo-Confucian Chu Hsi (whom we shall meet over his attacks on Buddhism, in chapter 14) only differed from Western materialism in that in the West matter obeys its own laws, which are unethical, while in Neo-Confucian China the material was subjected to laws that were ethical. Yet even though Neo-Confucian materialism was not mechanical – an atomic 'billiard ball' cause-and-effect outlook was quite foreign to it – this is still a mistaken view. What the Neo-Confucians realised was that the moral was fundamentally planted in Nature; it arose out of Nature, doing this by a sort of emergent evolutionary process and appearing, as we should put it, when the right conditions were present. The Neo-Confucians, then, approximated quite closely to the world-concept of an evolutionary materialism and to the philosophy of organism. On this view it would seem that the Neo-Confucians borrowed less from Buddhism than from Taoism, with which it seemed to join hands. This we shall now look at in more detail.

The Neo-Confucians

Chu Hsi and his predecessors. Between the Thang Taoists and the Sung Neo-Confucians there were a number of transitional figures, not necessarily earlier in time than the Neo-Confucians, but more impregnated with the Taoist tradition. They were not isolated, however, and at least one of them, Shao Yung (1001–77), was a friend of the chief leaders of the Neo-Confucian school.

Shao Yung, who in characteristic Taoist fashion refused all official responsibilities, turned out to be an original if fanciful thinker. His *Huang Chi Ching Shih Shu* (Book of the Sublime Principle which governs all Things within the World) was partly composed of elaborate diagrams in which cosmological and ethical ideas intermingled, but these were found so hard to understand that they were replaced by a descriptive account from the hands of his son with the help of another philosopher. Shao Yung also wrote an interesting philosophical dialogue, the *Yü Chhiao Wên Tui* (The Conversation of the Fisherman and the Woodcutter), in which he spoke, among other things, of the uniformity of natural phenomena. He retained the Tao as the name for the universal principle of Nature, but also gave a very high place to 'number' which he interpreted in a semi-mystical sense; the Tao first makes numbers, then 'forms', and finally fills these with matter. The two manifestations of the Tao were, he thought, motion and rest, motion generating the Yin and the Yang, rest being responsible for two entities that he himself introduced – softness and hardness. Earth was a mixture of these last two, while Heaven was a mixture of Yin and Yang. Each of the four entities could exist in two qualities, strong or weak, and from these eight possibilities Shao Yung derived every kind of phenomenon. True, he seems to have had a strong tendency to metaphysical idealism, and certainly his methods may have been fanciful and rather antique in style, but his was a very concrete and physical world-picture. Motion and rest, and to a lesser extent softness and hardness, were important ideas, and were taken over as basic concepts by the Neo-Confucians.

Shao Yung fully believed in the Indian idea, introduced with Buddhism, that there were periodic world catastrophes in which the cosmos dissolved into chaos and was then remade anew. The mediaeval concept of macrocosm and microcosm was also very vivid to him, but there is no evidence that this had been imported. Perhaps Shao Yung's most interesting concept from the point of view of the history of science was embodied in his expression *fan kuan*, or 'objective observation'. In science, he said, there are often things one cannot understand, and one must never attempt to force them into a scheme, because doing so brings in personal interpretations and prejudices, and then it is only too easy to fall into the error of making quite artificial constructions. Unfortunately Shao Yung hardly bore this in mind when he formulated his own theories. He believed, too, in a general community of observers: we are not restricted, he said, to personal observations, but we can use the eyes of all men as our eyes, their ears as our ears, and thus form a connected body of understanding about Nature.

Another philosopher who was also a transitional figure was Chhêng Pên, the author of *Tzu Hua Tzu* (Book of Master Tzu-Hua), who concealed his identity under the name of a philosopher of the Chou period. The text of this book is short and not very clear, though in spite of this it does give us some insight into the Taoist discussions then going on. It speaks of empty space, in which there are no barriers to the movement of bodies, and of equilibrium, in which there is no tendency for bodies to move in any direction. It also makes frequent references to fundamental forces, rhythms, or impulses, although none are very clearly explained. The writer attributed geometrical forms to the Five Elements; Water was straight, Fire pointed, Earth round, Wood curved, and Metal square. This was rather reminiscent of Kepler six centuries later associating the planetary spheres with the five figures of solid geometry, and it left its mark in the small pagodas or *chuang* characteristic of Buddhist temple gardens in China and Japan today. Admittedly there is nothing in his text to show that Chhêng Pên is thinking of particles here, but if he were not it is hard to imagine what he had in mind. His system also contains identifications of the elements with the organs of the body, and gives other hints about physiology and pharmacology.

We must turn now to the main school of the Sung Neo-Confucian philosophers. It had five leading personalities, who lived at various overlapping periods spanning the eleventh and twelfth centuries almost exactly. To put them into some sort of historical perspective, one should remember that the first four were contemporaries of the famous Arab poet, philosopher and scientist 'Umar al-Khayyāmī (Omar Khayyam), and the physician–scientist Ibn Sīnā (Avicenna), while the fifth and greatest Neo-Confucian lived at the time of the Arab physician–

philosopher Ibn Rushd (Averroes) and the Italian scholar and translator Gerard of Cremona. The Neo-Confucians were therefore working at about the same time as the peak of the movement in Europe that brought in translations of, and commentaries on, classical Greek works: they made their great synthesis of Confucian, Taoist and Buddhist ideas just before Thomas Aquinas, Europe's greatest synthesiser of Greek and Christian philosophy, began his career. If this is nothing more than a coincidence, it is certainly a remarkable one.

The five leading personalities of the Sung Neo-Confucian school were, first, Chou Tun-I (1017–73), a scholar who preferred philosophical study to high rank, and two younger men Chhêng Hao (1032–85) and Chhêng I (1033–1107), sons of a friend of his, both of whom attained great renown as philosophers. The fourth member of the group was Chang Tsai (1020–76), uncle of the two younger men, and he it was who seems to have been mainly responsible for the introduction of the acceptable elements from Taoism and Buddhism. Lastly there was Chu Hsi (1131–1200), the supreme synthesiser in Chinese history. Exactly how far he went in the study and practice of Taoism and Buddhism we do not know, but he was certainly learned in both systems and frequently refers to their doctrines, some aspects of which he incorporated in his own great philosophical synthesis. His official career was a very chequered one, periods of imperial favour alternating with resignations, retirement and deprivation of honours. Enormous literary output, unusual lucidity of expression, and an unswerving fidelity to a clear and definite world-picture, as well as great ability at organising the research and writing of others, place him without doubt among the greatest men in the whole development of Chinese thought. He has been likened to Aristotle, Thomas Aquinas, Leibniz and Herbert Spencer, among others, and it is a tribute to him that such suggestions seem in no way absurd. Perhaps Thomas Aquinas and the great Victorian philosopher Herbert Spencer, who both synthesised knowledge on the basis of scientific and social phenomena, are his closest equivalents, Aquinas because Chu Hsi was essentially a man of the mediaeval age, occupied with systematisation rather than transforming or superseding beliefs with a long history behind them; Spencer because Chu Hsi uncompromisingly affirmed a thoroughly naturalistic view of the universe, even though he lacked the vast background of assured experimental and observational evidence that Spencer inherited. It is a most striking feature of Chinese civilisation that its greatest philosopher should have been an Aquinas with a world-view like Herbert Spencer's.

The Supreme Pole. Chou Tun-I was more of a teacher than a writer, and left little behind. His fame mainly rests upon a very short exposition of a cosmical diagram, the *Thai Chi Thu Shuo* (Explanation of the Diagram of the Supreme Pole), a book on which Chu Hsi, taking its thesis as fundamental for his thought a century later, wrote a number of commentaries with similar titles. Figure 28 shows Chou Tun-I's diagram. His exposition is given below:

'(1) That which has no Pole! And yet [itself] the Supreme Pole! [*Wu chi erh thai chi*].

(2) The Supreme Pole moves and produces the Yang. When the movement has reached its limit, rest [ensues]. Resting, the Supreme Pole produces the Yin. When the rest has reached its limit, there is a return to motion. Motion and rest alternate, each being the root of the other. The Yin and the Yang take up their appointed functions, and so the Two Forces are established.

(3) The Yang is transformed [by] reacting with the Yin, and so Water, Fire, Wood, Metal and Earth are produced. Then the Five *Chhi* diffuse harmoniously, and the Four Seasons proceed on their course.

(4) The Five Elements [if combined, would form], Yin and Yang. Yin and Yang [if combined, would form], the Supreme Pole. The Supreme Pole is essentially [identical with] that which has no Pole. As soon as the Five Elements are formed, they have each their specific nature.

(5) The true [principle] of that which has no Pole, and the essence of the Two [Forces] and the Five [Elements], unite [react] with one another in marvellous ways, and consolidations ensue. The Tao of the heavens perfects maleness, and the Tao of the Earth perfects femaleness. The Two *Chhi* (of maleness and femaleness), reacting with and influencing each other, change and bring the myriad things into being. Generation follows generation, and there is no end to their changes and transformations.

(6) It is Man alone, however, who receives the finest [substance] and is the most spiritual of beings. After his [bodily] form has been produced, his spirit develops consciousness; [when] his five agents are stimulated and move, [there develops the] distinction between good and evil, and the myriad phenomena of conduct appear.

(7) The sages ordered their lives by the Mean, by the

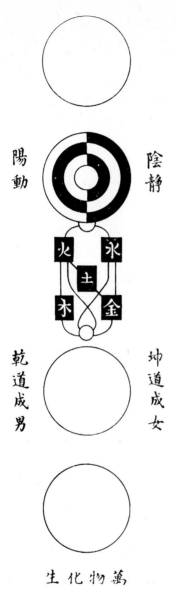

Fig. 28. The 'Diagram of the Supreme Pole' (*Thai Chi Thu*) of Chou Tun-I (A.D. 1017 to 1073). The second circle from the top is marked, on the left, 'Yang, motion', on the right, 'Yin, quiescence'. Below are the five elements. The second circle from the bottom is marked, on the left, 'The Tao of *Chhien*, perfecting maleness', on the right, 'The Tao of *Khun*, perfecting femaleness'. Below the lowest circle is written 'The myriad things undergoing transformation and generation'.

Correct, by Love and Righteousness. They adopted ataraxy as their dominant attitude, and set up the highest possible standards for mankind. Thus it was that "the virtue of the sages was in harmony with that of Heaven and Earth, their brightness was one with that of the Sun and Moon, their actions were one with the Four Seasons, and their control over fortune and misfortune was one with that of the gods and spirits." [A quotation from the *I Ching*.]

(8) The good fortune of the noble man lies in cultivating these virtues; the bad fortune of the ignoble man lies in proceeding contrary to them.

(9) Therefore it is said, "In representing the Tao of Heaven one uses the terms Yin and Yang, and in representing the Tao of Earth one uses the terms Soft and Hard, while in representing the Tao of Man, one uses the terms Love and Righteousness." And it is also said, "If one traces things back to their beginnings, and follows them to their ends, one will understand all that can be said about life and death."

(10) Great is the [Book of] Changes! [Of all descriptions] it is the most perfect.'

Chu Hsi explains:

'(a) The uppermost figure represents that of which it is said, "That which has no Pole! And yet [itself] the Supreme Pole!" It is the original substance of that motion which generates the Yang [force], and of that rest which generates the Yin [force].

It should be regarded neither as separate from, nor as identical with, the Two Forces.

(b) The concentric circles in the second figure symbolise motion giving rise to Yang and rest giving rise to Yin. The complete circle in the centre symbolises the substance which does this [equivalent to the circle in the first figure]. The semicircles on the left indicate the motion which produces Yang; this is the operation of the Supreme Pole when moving. The semicircles on the right indicate the rest which produces Yin; this is the substance when at rest. Those on the right are the foot from which those on the left are produced, and *vice versa* [i.e. Yang generating Yin, and Yin generating Yang].

(c) The third figure symbolises the transformations of the Yin and Yang forces in union with each other, and thus the generation of the Five Elements. The diagonal line from left to right symbolises the transformations of the Yang, and that from right to left symbolises the unions of the Yin.

Water is predominantly Yin and its place is therefore on the right. Fire is predominantly Yang and its place is therefore on the left. Wood and Metal are modifications [literally "tender shoots"] of the Yang and the Yin respectively, and therefore they are placed to the left and right under Fire and Water. Earth is of a mixed nature, therefore it is placed centrally. The crossing of the lines above the positions of Fire and Water indicates that the Yin generates Yang and *vice versa*. [The order of their generation is indicated by the intersecting lines connecting the Five Elements], Water being followed by Wood, Wood by Fire, Fire by Earth, Earth by Metal, and Metal again by Water, in an endless unceasing round, so that the Five *Chhi* spread abroad and the four seasons revolve.

(d) The Five Elements all come from the Yin and Yang [Forces]. The five different things [fit in to] the two realities without the slightest excess or deficiency. And the Yin and the Yang [go back to] the Supreme Pole [perfectly], neither one of them being more or less elaborate than the other, nor more or less fundamental than the other.

The Supreme Pole is essentially the same as that which has no Pole. Noiseless, odourless, it exists everywhere in the universe. As soon as the Five Elements are generated, they have each their specific natures. Since these *chhi* are different, the tangible matter [which manifests them] is also different. Each sort has its completeness, and this there is no gainsaying.

The small circle below, connected by four lines with the Five Elements above, indicates that which has no Pole, in which all are mysteriously unified, as indeed again cannot be denied.

(e) The fourth figure represents [the operations of the *chhi* of Yin and Yang exhibited in] the principles of [heavenly] maleness and of [earthly] femaleness [which pervade the universe], each having their own natures, but [both going back to] the one Supreme Pole, [as indicated by the reproduction of the original circle].

(f) The fifth figure represents the birth and transformation of the myriad things in their sensible forms, each of which has its own nature. But [as indicated again by the reproduction of the original circle], all the myriad things go back to the one Supreme Pole.'

The most amazing statement in Chou Tun-I's credo is that of his first paragraph. It is essentially a statement of a synthesis uniting streams of Taoist and Confucian thought. Chu Hsi confirms this. The word *chi* that appears in Chou Tun-I was from old the technical term for the poles of the heavens, so in essence we have a statement saying that around the Pole Star all man's universe revolved. Concerned as Chou Tun-I and Chu Hsi were with the conception of the entire universe as an organism, the *chi*, the pole, was a kind of organisational centre, the very world-axle itself. Yet there were also two opposites to be reconciled here, for the Taoist words *wu chi* were an affirmation that the true and entire universe depended on no such cardinal point, since every part of the organism took its leadership in turn, while the Confucian *thai chi* were a recognition of an inherent power present everywhere in the universe, a universal process. Thus Chou Tun-I and Chu Hsi had arrived at the idea of the world as indeed a single organism, no particular part of which can be identified as in control.

Modern minds have become accustomed to thinking (or consciously not thinking) in these terms; the world is full of poles and centres, magnetic fields, cells and their nuclei, centres of social control in times of peace and war. But they are all secondary to the organisms of which they form part, not superior to them. To the Neo-Confucians the world was no less differentiated than it is for us: it manifested itself at various levels of organisation, things that are whole at one level of organisation being parts on the next – a chemical compound at one level being separate elements at another, and (to us) separate atomic particles at yet another. Though entirely natural, the highest human values are only applicable at the human level. Chu Hsi is quite specific on this point, and in Neo-Confucian writings we even find a technical term for the concept 'level of organisation'. In brief, then, the identity of *thai chi* and *wu chi* was the recognition of two things – the existence of a universal pattern determining all states and transformations of matter and energy, and the presence of this pattern everywhere. The motive power could not be localised at any particular point in space and time. The organisation centre was identical with the organism itself.

When one takes a further look at this pithily expressed system of Nature, one cannot but admit that the Sung philosophers were working with concepts not unlike some being used in modern science. No doubt the idea of two fundamental forces was an ancient generalisation based on the two sexes of so many species, man included, and the Sung thinkers may have done little more than tabulate its consequences; yet the more one reads them, the more one feels that they really did attain an inkling of those two deeply rooted aspects of matter which appeared later in the West as positive and negative electricity, as protons and electrons, as the components of all material particles. It was something they could not express, but it was nevertheless a true insight. Here the Chinese, without ever attaining the position of Newton, shot an arrow close to the spot where the atomic physicists Bohr and Rutherford were later to stand.

Another deep conviction which clearly emerges from Neo-Confucian works is that Nature worked in a wave-like manner. Each of the two forces rose to a maximum in turn and then fell away, leaving the field to its opposite; moreover, they generated each other in a way reminiscent of that 'interpenetration of opposites' favoured by some modern philosophers. The constant references to motion and rest, occurring in alternate periods, the motion rising to a maximum and then returning to a zero point, expresses what we should consider a quite legitimate scientific abstraction of wave-like phenomena.

Again, in the quotations given, there is a rather clear notion of the production of new things by reactions that we can almost term chemical, and in one place, indeed, an alchemical symbol is actually used. Although it is frequently said that before Neo-Confucianism China had no real metaphysics, we can only remark that if it was they who introduced it, then it was a type of metaphysics very much in harmony with physics in the modern scientific sense.

Chou Tun-I's other work, the *I Thung Shu* (Fundamental Treatise on the Book of Changes), is at first sight wholly concerned with ethical matters remote from the natural sciences – the sage and his rôle in society, his wisdom, and matters like rites, music, and so on – and the argument of the book centres on the technical term *chhêng*, the normal meaning of which is 'sincerity'. However, from the much earlier *Chung Yung* (Doctrine of the Mean), a book of the Chou dynasty, we find a punning reference which runs: 'He who is sincere [*chhêng*, 誠], perfects [*chhêng*, 成] himself', and this gives the clue that *chhêng* is not only 'sincerity'; it is a quality of something capable of being inherent in an individual, and not only something arising from relations between individuals. It is therefore more what we might call 'integrity' than 'sincerity' – being sincere with oneself, not deluding oneself nor acting contrary to one's true nature. The *Chung Yung* also says 'Sincerity is the Tao of Heaven; to apply oneself to sincerity is the Tao of Man', indicating that it transcends the human sphere. Heaven has *chhêng* because it faithfully follows its true nature and does nothing against its Tao. Thus we come to the realisation that *chhêng* is achieved when every organism fulfils with absolute precision whatever its function may be in that higher organism of which it is a part, and here again we have a state of affairs that is quite familiar to modern philosophies of organism.

The cosmic significance of the *chhêng* of the *Chung Yung* was developed fifteen centuries later in many places in the *I Thung Shu*. Just as the first *kua* in the Book of Changes symbolises the beginning of all things, so it is the origin of *chhêng*, which comes into being with them; it is pure, perfect, exerts no force, it is (or it generates) all good and evil and, like the Tao, it has a virtue which transmutes private love into universal human-heartedness, rightness into righteousness, natural human patterns into social order. Diffusing outwards, it evokes beginnings and developments; ebbing, it leaves permanent gains. When at rest it is as if it did not exist; when in action its existence is manifest. Like the individual patterns which orient themselves to its influence, it belongs to the category of unseen (spiritual) things in the universe. Once more we have a concept that could fuse together, in an evolutionary scheme, the natural world of the Taoists and the moral world of the Confucians: a system of philosophical naturalism with great relevance to the natural sciences.

Of the other members of the Neo-Confucian School, Chhêng Hao tended towards metaphysics, and his brother Chhêng I made several worthy scientific points, yet neither is especially relevant to those aspects of Sung philosophy with which we are concerned. Their uncle, Chang Tsai, however, did pay particular attention to one aspect which we shall now be constantly meeting – the formation of all things and all living creatures by a process of dispersion. The same technical terms had been used by the sceptical philosopher Wang Chhung (Chapter 11) a thousand years earlier, but for Chang Tsai, like the other Neo-Confucians, the world contained nothing supernatural.

The conception of the formation of things by the aggregation of the universal 'matter-energy', *chhi*, was fully taken over by Chu Hsi, who said quite specifically that the *chhi* condenses to form solid matter. But what he did that was new was to associate the aggregation process with the Yin, and the dispersion process with the Yang, and after his time, from the Sung onwards, these doctrines of expansion and condensation became part of the fabric of Chinese thought.

The study of the universal pattern. We must now turn and look more closely at the naturalist philosophy of Chu Hsi. He worked with two fundamental concepts, the *chhi* (氣) and the *Li* (理), which he took as representing the material and non-material elements of a basically naturalistic universe. The significance of Chu Hsi's *chhi* cannot, however, be expressed by one single English word: it could be a solid or a gas or vapour, but equally well an influence as subtle as those which phrases like 'electro-magnetic waves' or 'fields of force' have for modern minds. Sinologists themselves generally translate Chu Hsi's use of *chhi* simply as 'matter', yet he uses another term *chih* for matter in its solid, hard and tangible state. Though *chih* is a form of *chhi*, *chhi* is not always *chih*, because matter can exist in tenuous non-perceptible forms. Perhaps, then, 'matter-energy' is a better translation, if one must be used.

There has been much disagreement over *Li*. An early tendency was to translate it as 'form', but this has links with Aristotle and Greek philosophy which are certainly not relevant. Nor should it be translated as (scientific) 'law', since this would prejudge the question of whether or not the Chinese at any time developed the ideas of 'laws of Nature'. Other equally unacceptable translations have been suggested, but from the point of view of the scientist, at any rate, it may be best to think of *Li* in terms of organisation, since the universe can be described in terms of organised levels. For instance, one level of organisation will give the configuration of an atom, another (higher or coarser) level, that of a molecule, another again a living cell, which is, of course, an envelope composed of many lower-level organised spatial envelopes one within the other. This is not to suggest that Chu Hsi and his Neo-Confucian colleagues talked like this, or even implied these ideas in detail, so it would be dangerous to use a phrase such as 'levels of organisation' in translation. But since that concept is inherent in their use of *Li*, the most satisfactory course is to translate *Li* as 'pattern', and hence by the single word 'organisation', for Chu himself consciously applied *Li* to include the most vital and living patterns known to Man, and something of the idea of 'organism' was what really lay at the back of the Neo-Confucians' minds. We could then translate *chhi* as 'matter-energy' and *Li* as 'organisation' or 'principles of organisation', and these terms will then show us that in one sense the Neo-Confucians were not very far from a view of the universe similar to that of the modern natural scientist and organic philosopher. On this reckoning Chu Hsi, who removed *Li* from its Buddhist context and restored it to its ancient naturalistic place, was more advanced than any of his interpreters and translators, Chinese and European, have yet given him credit for.

It is now time we listened to Chu Hsi himself. In his collected works we read:

'Throughout Heaven and Earth there is *Li* and there is *chhi*. *Li* is the Tao [organising] all forms from above, and the root from which all things are produced. *Chhi* is the instrument [composing] all forms from below, and the tools and raw material with which all things are made. Thus men and all other things must receive this *Li* at the moment of their coming into being, and thus get their specific nature; so also they must receive this *chhi*, and thus get their form.'

So far the text clearly justifies the interpretation of *chhi* as matter-energy and *Li* as the cosmic principle of organisation at all levels. Yet was there any precedence or priority between them? Chu Hsi was a little hesitant:

'First there was *Li* and later there was *chhi*. This is what the *I Ching* means when it says "One Yin and one Yang go to make the Tao . . ."

'First there is the *Li* of Heaven, then there is the *chhi*. The *chhi* agglomerates to form the *chih* and that is the preparatory raw material for the "nature".

'Someone asked whether *Li* or *chhi* came first. The philosopher answered, "*Li* is never separated from *chhi*. But *Li* is above all form [non-material] while *chhi* is below all form [material]. If one has to speak of above and below in this way, there could hardly but be a before and after. *Li* has no form, but *chhi* is gross and contains [impure] sediments.

'"Yet one cannot really speak of any priority or posteriority of time as between *Li* and *chhi*; it is only if one insists on considering their origins that one has to say that *Li* came first. *Li* is not some kind of separate thing, it has [necessarily] to inhere in *chhi*. If there were no *chhi*, *Li* would have no way of manifesting itself and no dwelling-place. *chhi* can produce the Five Elements, but *Li* can produce [also] Love and Righteousness, Good Customs and Wisdom . . ."

'Someone else objected, saying, "You speak of *Li* as first and *chhi* as second, but it seems that one cannot apportion to either of them priority or posteriority." The philosopher replied, "I do wish to retain a sense in which *Li* is first [and *chhi* second]. But you can never say that here and now is *Li* while tomorrow there will be *chhi*. And yet there is [in some sense] a before and after."

'Someone asked yet again whether *Li* was prior and *chhi* posterior, and the philosopher replied, "Fundamentally one cannot say that there is any difference between them in time, but if one goes back in thought to the beginning of all things, one cannot help imagining that *Li* was first and *chhi* came after."'

Clearly Chu Hsi was trying to avoid falling into the pit of idealism, but he did not wish to give the impression of being a materialist either. He was anxious to avoid being pushed into saying matter-energy arose from organisation or *vice versa*. He nevertheless inclined to the former view, partly because it was so difficult to think of organisation as a category independent of mind, and to get rid of the idea that a plan implies a planner who must be prior in time and superior in status to whatever is planned. At bottom, Chu Hsi remained a dualist in the sense that *chhi* and *Li* were coaeval and of equal importance in the universe, 'neither afore nor after other', though the residue of a belief in some slight superiority of *Li* was extremely difficult to discard. Presumably the reason for this was unconsciously social, since in all forms of society which the Neo-Confucians could think of, the administrator in charge of planning, organising and administrating was always socially superior to the farmer and the artisan who were occupied with material things, with matter, and so with *chhi*. If Chu Hsi could have liberated himself from this prejudice, then he would have anticipated by eight hundred years the standpoint of organic materialism.

All this was bound to be reflected in Chu Hsi's theory of the nature of knowledge. Cognition or apprehension is, he thought, the essential pattern of the mind's existence, but that the mind can do this is evidence of what we may call the 'spirituality inherent in matter'. In other words, the mind's function is perfectly natural, something which matter has the potentiality of producing once it has formed itself into arrangements with a sufficiently high degree of pattern or organisation. As to the principle of organisation itself, it is *Li* which prevents the process of Nature from falling into confusion:

> 'Take, for example, the Yin and the Yang and the Five
> Elements; the reason why they do not make mistakes in their
> counting, and do not lose the threads of their weaving [i.e. do
> not fall into irremediable disorder], is because of *Li*.'

Li is identified with the Supreme Pole, of which everything has its share: without it nothing could have come into existence. It is connected with the properties of motion and rest – the modern concepts of energy and inertia, corresponding to the active Yang and passive Yin and, indeed, to the appreciation of motion in the universe:

> 'After the heavens and the Earth were formed, it was the
> active principle [*Li*] which imparted to them their gyratory
> movement. Each day has its diurnal revolution and each
> month and year their revolution [of the heavenly bodies] . . .
> 'The heavens revolve without resting. Dawn and night
> revolve as if on well-polished bearings. The Earth is like a

bridge in the middle. If the heavens stopped for a single instant, the Earth would fall to destruction.'

To Chu Hsi, *Li* also had an interesting relationship to mathematics:

'Someone asked about the relation of *Li* to number. The philosopher said, "Just as the existence of numbers follows from the existence of *Li*, so the existence of numbers follows from the existence of *chhi*. Numbers, in fact, are simply the distinction of objects by delimitation."'

Here was the germ of something that could have revolutionised Chinese science – the missing mathematical approach to theories of Nature. But it is only a momentary flash; we hear no more about it, and it is to be feared that the 'numbers' referred to were those of sterile Pythagorean numerology (see p. 159), not a form of mathematics that would have been of help to natural science.

With Chu Hsi and other Neo-Confucians there was a relation between *Li* and Tao. Tao was the pattern of all things, *Li* the pattern inherent in any natural object. 'The term Tao', Chu Hsi says, 'refers to the vast and great, the term *Li* includes the innumerable vein-like patterns included in the Tao.' Thus 'Tao' was to be used for the cosmic organism, while *Li* could be used also for the minutest patterns of small individual organisms. But in line with the Confucian tradition, which the Neo-Confucians could not desert, 'Tao' was used more often for the Tao of man in human society than for the Tao of non-human Nature. Yet by his doctrine of *Li* and *chhi* Chu Hsi managed to bring together the two uses of 'Tao' – by the ancient Confucians and the ancient Taoists – into one scheme: the Tao of human society was seen to be that part of the Tao of the cosmos which makes itself evident at the organic level of human society. In this way the two great schools of Chinese thought attained a real synthesis.

Evolutionary naturalism. The idea that the universe passed through alternating cycles of construction and dissolution was common ground for all the Neo-Confucians. It seems to have been considered in a systematic way by Shao Yung, who began by applying the twelve hours and the twelve points of the Chinese compass to its different aspects, and he was followed by other Neo-Confucians. For example Wu Lin-Chhüan (1249–1333), who lived in Yuan times, wrote:

'The cosmic period is one of 129600 years, divided into 12 *hui* of 10800 years each. When Heaven and Earth, in their revolutions, attain the eleventh *hui*, all things are closed

down, and all men and beings between Heaven and Earth
come to nothingness. After 5400 years the position of *hsü* is
past, and when the middle of the twelfth *hui* is reached, the
heavy and gross matter which, in solidifying, had formed the
Earth, becomes dissipated and rarified, joining with the
tenuous matter which had formed the heavens, and uniting in
one single mass; this is called Chaos. This mass then acquires
an accelerating rotational movement, and when the position
hai is coming to its end, the material reaches its darkest and
most dense condition.

'At the point *chêng*, the Great Period begins again and a
new era opens; it is the beginning of the first *hui*. . . Thence-
forward, light gradually increases. After another 5400 years. . .
the lightest part of the mass separates and rises, forming sun
and moon, planets and fixed stars. . .

'When the middle of the second *hui*. . .is reached, the
heaviest *chhi* condenses forming rocks and earth, and its
liquid part becomes water. . .while its calorific part becomes
fire. . .

'Another 5400 years, and the middle of the third *hui*. . .is
reached, and now human beings begin to be born between
Heaven and Earth.'

One cannot but see in descriptions like this the Chinese predilection for
wave-forms, as in the alternating dominance of Yin and Yang. In one
sense they were unsupported speculations, although it would be going
too far to say that they did nothing for Chinese science. Apart from their
naturalism, they helped the Chinese to arrive at advanced notions of
geology and to recognise the true nature of fossils much earlier than was
the case in Europe. Chu Hsi distinctly mentions these matters himself;
echoes of discussions about the exact lengths of times involved appear
in the *Chu Tzu Chhüan Shu* (Complete Writings of Master Chu), while
ideas about the solidification of rocks deep underground and the
formation of some in water (sedimentary rocks) were debated.

Chu Hsi also had views about the origin of life. He believed that
spontaneous generation – the production of living organisms from non-
living material – had once played a great part in producing life, and was
still active to some extent:

'At the beginning of the generation of beings, the most subtle
parts of the Yin and the Yang condensed to form two
[components], like the spontaneous appearance of lice, which
burst forth [under the influence of warmth]. But when two

individuals, one male and one female, had been brought into
being, their succeeding generations came from seeds, and this
is the most universal process.'

As to the nature of the lower animals, he clearly recognised that the
categories and values applicable to human society could not be applied
to them. The behaviour of social insects like ants and bees shows a 'gleam
of righteousness', that of mammals and their offspring a 'gleam of love',
but animals have a material constitution which is gross and opaque –
a lower level of neurological organisation, as we might now say – and
this prevents them from realising the full possibilities of Nature. Animals
behave as they do, not by choice, but because of the Tao or *Li* which
they have to follow. Thus when consciousness appears at the human
level it is not wholly unconnected with Man's material composition.

Modern science could find little to quarrel with in these views which,
be it remembered, date from the middle of the twelfth century. To Chu
Hsi, as to us, the highest human virtues are natural, not supernatural;
the highest manifestations of the evolutionary process. Chu Hsi also
spoke of love (the principle of aggregation in the universe) as the motive
force of all things. Man, being endowed with *chhi*, receives the mind
of Heaven and Earth, and thereby his life; his tender-heartedness and
love are part of the very essence of this life. Moreover, just as transition
from the lower to the higher animals depends on the relative purity of
their *chhi*, so also the differences in goodness and badness between
human beings depends on the inequality in their endowment of *chhi*.
But Chu Hsi does not develop this fatalistically as Wang Chhung did;
Chu claims that Man can improve by using the *Li* within himself. These
views are all of considerable interest, for in many ways they
foreshadowed modern genetics.

Chu Hsi's thinking about 'organism' in its wider sense is present
throughout all he writes on human and other relationships. It comes
out particularly clearly when, in opposition to the Buddhists, he is
discussing the nature of social organisation:

> 'Under heaven, only the principles of Tao and *Li* exist, and
> we cannot but follow them unto the end. The Buddhists and
> Taoists, for example, even though they would destroy the
> social relationships [i.e. by becoming monks and cutting
> themselves off from the world], are nevertheless quite unable
> to escape from them. Thus lacking [the relationship of] father
> and son, they nevertheless pay respect to their own preceptors
> [as if they were fathers] on the one hand, while they treat
> their novices as sons on the other . . .'

Table 16. *Rationalisation of Confucian terms by the Neo-Confucians*

Associated with Yang 陽	Associated with Yin 陰
chhi 氣 used in its old sense as the 'breath of life'	*ching* 精 the seminal essence
hun 魂 the 'warm' part of the spirit or soul, which at death ascends to mingle with the *chhi* of the heavens	*pho* 魄 the 'cold' part of the spirit or soul, which at death descends to mingle with the *chhi* of earth
shen 神 the ancient term for a god, now used to express the concepts of:	*kuei* 鬼 the ancient term for a demon, now used to express the concepts of:
shen 伸 expansion, disaggregation and	*chhü* 屈 contraction, aggregation and
san 散 dissipation, dispersion	*chü* 聚 collection, condensation

Chu Hsi also talked of life and death, and was quite certain that individual human spirits did not survive as such. The Buddhist opinion that human souls survive as ghosts and are reincarnated in other human beings is absolutely wrong: the only changeless thing in the universe is *Li*. In this connection Chu Hsi and the Neo-Confucians rationalised the old Confucian terms into terms of different technical meaning. We can draw up a table (Table 16) showing these changes, and although there was nothing new here in the idea that the human souls were composed of two groups, one ascending and the other descending at death, the Neo-Confucian innovation was to give these terms rather clear physical expression, and then to apply them in describing natural phenomena. Thus Chu Hsi:

> 'When wind, rain, thunder and lightning occur, this is the operation of *shen* [gods, i.e. expansive forces]. When the wind goes down, the rain ceases, the thunder is ended, and the lightning flashes no more, this is the operation of *kuei* [demons, i.e. contractive forces].'

Such a statement can, of course, be taken two ways – one superstitiously by the mass of people, using demons as in folk religion, the other naturalistically as adopted by the scholar – but this ambivalence must be seen against the background of Chinese bureaucratic society, where ceremonies like praying for rain were still carried out, even though the enlightened did not believe they had any effect at all. But perhaps, in retrospect, it may be that the failure to introduce new technical terms, with the tendency merely to rationalise the old ones with all their religious undertones, was one of the more unfortunate aspects of the social milieu in which Chinese science had its birth.

We come finally to the question of the existence of God in the Neo-Confucian outlook. Chu Hsi's position was unequivocal:

> 'The blue sky is called Heaven; it revolves continuously and spreads out in all directions. It is now sometimes said that there is up there a person who judges all evil actions; this assuredly is wrong. But to say that there is no ordering [principle] would be equally wrong.'

His standpoint fixed Confucian orthodoxy and, later on (Chapter 16), we shall enquire how far this view really contributed to the development of a scientific world-outlook in China.

Neo-Confucianism and the golden period of natural science in the Sung

From what has been said, it seems that to classify the philosophy of the Neo-Confucians as a philosophy of organism is valid enough, but it is worth while emphasising again that the period of the Sung dynasty was precisely that which saw the greatest flowering of indigenous Chinese science. As we saw earlier, Taoism might be expected to have had some connections with practical science, and this was so; alchemy, the use of plant-drugs, the classification of animals and the physics of magnetism were all Taoist in inspiration. So also, if the views here expressed about Neo-Confucian philosophy are correct, it should show this connection too, and that indeed was the case. A large amount of evidence comes forward to prove the point.

If, for instance, we run over the great scientific names of the period, we find that Shen Kua was the first to describe in detail the magnetic compass, to speak of relief maps, and to study fossils and recognise their nature. In mathematics there was a host of names – Liu I, Li Yeh, Chhin Chiu-Shao and Yang Hui, to mention only the few who worked out that Sung algebra which constituted the most advanced mathematical skill anywhere in the world at that time. In astronomy there was Su Sung, whose elaborately illustrated book of 1086 on the armillary sphere and clock still exists; while a century and a half after his time, the famous Suchow planisphere or chart of the heavens was carved on stone. In geography and quantitative map-making a new era opened with Chia Tan and culminated in Chu Ssu-Pên, both among the greatest geographers the world has ever seen. Many alchemical books were also written during the Sung, and in botany and zoology the output was extraordinary. In the Wu Tai (907–60) and Sung periods nine great treatises on pharmaceutical natural history came out, as well as a host of specialised monographs in botany. Medicine flourished too, and in Sung Tzhu (1247) we have the founder of forensic medicine not only in China, but in the whole world. Basic books on architecture and military technology were also produced at this time.

Neo-Confucian philosophy, essentially scientific in quality, was certainly accompanied by a hitherto unparalleled flowering of all kinds of activity in the pure and applied sciences. Yet none of this brought Chinese science to the level that it was to reach in the West with men like Galileo, Harvey and Newton. After a retardation in the Yuan (Mongol) and the Ming dynasties, and an upsurge of humanistic learning in the Chhing (Manchu) period, it becomes evident to us, as we look back, that barring some miracle, Chinese civilisation was not going to produce modern natural science. Instead, the last act of the drama was played out in terms of rather sterile metaphysical controversies. Only at the beginning of the seventeenth century, when the Jesuits, the first emissaries of post-Renaissance modern-Western science, reached the Chinese capital, were Chinese scholars invited to join in the 'new, or experimental, philosophy' that was to bring about a fundamental transformation of the world.

What the Jesuits contributed in bringing modern mathematics, science and technology from Europe to China was immense, but it would be a mistake to suppose that Chinese thought, summed up in Neo-Confucianism, gave nothing to world science: its contribution was probably far greater than has previously been recognised. The reason for this new assessment is simple enough: European science marched under the banner of a mathematical and mechanical universe, the world-picture of Descartes and Newton. This was a procession that carried all before it, but it espoused a view of Nature which could not permanently satisfy the needs of science: there was bound to come a time when it was necessary to look on physics as the study of the smaller organisms and biology as the study of the larger organisms. Once this had happened, science had then to draw on a mode of thinking that was very old, very wise, but not characteristically European at all.

What the Chinese contributed was this new element. The world-outlook of Chu Hsi and the Neo-Confucians stimulated an organic quality of thinking, a quality that came by way of the Jesuits to Europe and to Leibniz, a quality the importance of which can hardly be exaggerated. Leibniz was a key figure in this transfer. He had great interest in Chinese thought, and kept up a constant correspondence, even some personal contact, with the Jesuits centred in Peking. He can in some ways be said to have acted as a bridge-builder, spanning the gap between the idealism of theology and the materialism of European science, for there was a contradiction between these two that European thought had never succeeded in solving. Precisely how much stimulus Leibniz actually received from the Chinese is difficult to assess, since he was an unsystematic writer, and much of what he did remains only in the form of correspondence and notes. But something can be said.

Leibniz wanted a realism but not a mechanical one. Against the view of the universe as a vast machine, Leibniz proposed the alternative view of it as a vast living organism, every part of which was also an organism. This was the view he presented in 1714, at the end of his life, in a short but brilliant treatise, the *Monadology*, which was published post-humously. His 'monads' were indissoluble organisms participating as parts of higher organisms, and were organised on different 'levels': their hierarchy and their 'pre-established harmony' made them resemble the innumerable individual manifestations of Neo-Confucian *Li*. Thus Leibniz tried to bridge the idealistic-materialist gap in the West by using a universe of integrative levels, and there is little difficulty in finding elements of Chinese thought in his philosophy. In the *Monadology* he wrote: 'Every portion of matter may be conceived of as a garden full of plants or a pond full of fish; but every stem of a plant, every limb of an animal, every drop of sap or blood is also such a garden or pond.' Here we can see Buddhist speculation viewed through a Neo-Confucian glass, yet meeting with experimental verifications seen through the then novel microscope by men like Anton van Leeuwenhoek and Jan Swammerdam, to whom he admiringly refers.

When Leibniz speaks of the difference between machines and organisms as lying in the fact that every monad constituting the organism is alive and co-operating in a harmony of wills, we are irresistibly reminded of the 'harmony of wills' so characteristic of Chinese associative thinking. And although his 'pre-established harmony' – a doctrine designed to solve the mind–body problem as seen in the seventeenth century – has not lasted as such, it was a doctrine that echoes most closely the views quoted from Tung Chung-Shu (second century B.C.) (see p. 164), both using the analogy of sound-waves. Yet another echo of Chinese thought may be sensed in the passage where Leibniz says: 'There is neither absolute birth nor complete death in the precise meaning of the separation of soul and body. What we call births are developments and unfoldings, what we call deaths are foldings and shrinkages.' Here we can almost hear the Taoists talking about their condensations and rarefactions.

Leibniz' own opinions on Chinese philosophy still survive, a philosophy that he makes clear is 'modern' Chinese, i.e. Neo-Confucian. He wrote:

> 'Thus we may applaud the modern Chinese interpreters when they reduce the government of Heaven to natural causes, and when they differ from the ignorant populace, which is always on the look out for supernatural [or rather supra-corporeal] miracles, and Spirits like *Deus ex machina*. And we shall be able to enlighten them further on these matters by informing them of the new discoveries of Europe, which have furnished almost mathematical reasons for many of the great marvels of Nature, and have made known the true systems of the macrocosm and the microcosm.'

Of course no one would want to say that the stimulus of Chinese organic philosophy was the only one that led Leibniz to his new ideas. He found many points of contact between his own views and those of the 'Cambridge Platonists', the seventeenth-century theological-philosophical school led by such men as Benjamin Whichcote, Henry More and Ralph Cudworth, who, with their friends the biologists Nehemiah Grew and John Ray, developed a philosophy of science that saw Nature as 'vital', 'spermatical' and 'plastic', not mechanical. The Cambridge Platonists wanted an understanding and contemplation of Nature, not a control over it; they sought synthesis not analysis. But it was a view based to so great a degree on Plato that it was limited in what it could achieve, and its supporters turned their back on a mathematical universe in favour of a 'spirit of vitalism'. Yet in the seventeenth century the path away from Descartes and Newton did not turn away from mathematics: it passed directly through its midst. This is the path that Leibniz took, but it could only be taken in the light of an organic universe from which every trace of animism, of the soul as a vital principle of organic development, had been purged. Perhaps Neo-Confucian *Li* showed the way for this purification. And perhaps now, at our present stage in modern science, the world owes more to men like Chuang Chou, Chou Tun-I and Chu Hsi than it has yet realised.

13

Sung and Ming idealists and the last great figures of Chinese naturalism

After the death of Chu Hsi in 1200 there was little development of Neo-Confucianism. Some pupils and colleagues tried to apply his principles in specific fields, some elaborated special theories about the doctrine of universal catastrophes, and still others busied themselves with collecting and publishing the remains of Chu Hsi's work, but there was little real development of Neo-Confucianism as such. On the other hand, the predominant thinkers of the fourteenth, fifteenth and sixteenth centuries did concern themselves with *Li* and *chhi* and tried to reach some kind of ultimate unity between them. As Yang Tung-Ming (1548–1624) wrote:

> 'One can say that the nature of social values and *Li* arise out of energy-and-matter (*chhi chih*); but not that energy-and-matter arise out of social values and *Li*.'

Indeed Yang Tung-Ming and others of like mind were all in strong opposition to the tradition of metaphysical idealism – the assertion that reality lies only in our minds and not in the external world – which had come to a climax with Wang Yang-Ming about 1500, and this is the situation we must first examine. Only briefly, however, for idealism, whatever its form, was no more helpful to the sciences in China than in any other civilisation. But it became popular, and so it was yet another weight to put on the scales against Chinese science.

The responsibility for the upsurge in idealism seems to lie at the door of Buddhism. In ancient Chinese thought there is no evidence of idealist metaphysics, and its first real growth did not occur until the Thang dynasty, with the advent of Buddhist teachers such as Lu Hui-Nêng (630–713) and Ho Tsung-Mi (779–841); in fact it was essentially a development of the Indian philosophy of *māyā*, the basic unreality of the external world, and its greatest exponent was undoubtedly Lu Chiu-Yuan (1138–91), a contemporary and strong opponent

250

of Chu Hsi. Lu stated the doctrine with greater emphasis and precision than any of his predecessors:

> 'Space and time are [in] my mind, and it is my mind which [generates] space and time. . .
> 'The myriad things are condensed into a space, as it were, of a cubic centimetre, filling the mind. Yet emanating from it, they fill the whole of time and space.'

This could be seen as virtually an anticipation by six hundred years of Immanuel Kant's idea of the subjective nature of space.

Lu Chiu-Yuan placed the Neo-Confucian principle of *Li*, or organisation, wholly within the mind, and after years of argument with Chu Hsi the two men had to agree to differ, since their views were totally irreconcilable. Because of his opinions, it is not surprising that Lu Chiu-Yuan was accused of partiality towards Buddhism, and certainly, like many other late Confucians, he adopted various techniques of meditation, but there is no doubt that both he and his contemporaries always emphasised the duties of man in the world of affairs, and denied the Buddhist doctrine of salvation by escape from the world. Indeed, according to a statement current among Lu Chiu-Yuan's circle, 'Buddhist meditation leads towards extinction, Confucian meditation towards action.'

Lu Chiu-Yuan's influence extended through a succession of pupils down to the sixteenth century and thus to Wang Yang-Ming (1472–1528), the chief representative of Chinese idealism, who constantly described himself as a follower of Lu Chiu-Yuan. Wang Yang-Ming, however, expressed his idealism in different terms from his predecessors; in his selected works (*Yang-Ming hsien-sêng Chi Yao*) we find:

> 'The master of the body is the Mind; what the mind develops are Thoughts; the substance of Thought is Knowledge; and those places where the Thoughts rest are Things.'

For Wang Yang-Ming the external world was not less real than the world of the imagination, but all material objects were unquestionably the product of the thought of the 'world-spirit' with which the thoughts of every individual were, in some way or other, identical. Hence the great emphasis he placed on inborn 'intuition', apart from which there could, he thought, be no knowledge. Such intuition was often conceived of in a very ethical way, a kind of moral intuition, for which Wang Yang-Ming, as a true Confucian, believed he had the authority of the ancient sage Mencius. In many senses, then, he anticipated by two hundred years the philosophy of Bishop Berkeley, and by still longer

Kant's 'categorical imperative' – the existence of a moral law independent of any ulterior motive or end.

Wang Yang-Ming was a considerable poet as well as a philosopher, and some of his poetical writing, long commonplace in China, has by now become part of world literature; for instance:

> 'Everyone has a Confucius in his heart
> Sometimes visible but sometimes hidden,
> Without many words one may point to what it is –
> The innate knowledge of goodness which admits of no
>   doubts.'

For us it might be the 'inner light', the Johannine 'light that lighteth every man that cometh into the world'. Unfortunately all this, sublime though it was, could hardly be sympathetic to the development of natural science. What is more, Wang Yang-Ming could never understand the basic principle of scientific method. The 'objective investigation of things' (*ko wu*) was a phrase, much used by Chu Hsi, which had come down from antiquity, first appearing in the *Ta Hsüeh* (Great Learning) of 260 B.C. It was to be the watchword of Chinese scientists all through the ages. But this is how Wang Yang-Ming wrote about it:

> 'In former years I discussed this with my friend Chhien,
> saying, "If to be a sage or a man of virtue one must
> investigate everything under Heaven, how can anyone at
> present acquire such tremendous strength?" Pointing to some
> bamboos in front of the pavilion, I asked him to investigate
> them. So both day and night Chhien [sat and] investigated the
> principles of the bamboos. After three days he had exhausted
> his mind and thought, so that his mental energy was fatigued
> and he became ill. At first I said that this was because his
> energy and strength were insufficient, so I myself undertook to
> carry on the investigation. But day and night I was unable to
> understand the principles of the bamboos, until after seven
> days I also became ill, having been wearied and burdened by
> thoughts. Thus we both sighed and concluded that we could
> not be either sages or men of virtue, lacking the great strength
> required for carrying on the investigation of things. And,
> moreover, during the three years which I spent among the
> tribesfolk [Wang Yang-Ming was banished for some years] I
> found that no one could possibly investigate everything in the
> world. And I came to the conclusion that research could only
> concentrate introspectively on one's self. This leads to a
> wisdom within the reach of every man.'

The tradition enshrined here continued throughout the sixteenth and seventeenth centuries, but so far as science was concerned it had little relevance. Idealist doctrine was by then no longer in the ascendant; a great 'reverse' movement was under way, a new naturalism which was so vigorous that it even went so far as to criticise Chu Hsi for not having been materialist enough in his outlook.

The reaffirmation of materialism

One of the earliest and most eminent representatives of this new movement was Wang Fu-Chih (1619–92). An eminent scholar, he served the Ming dynasty as long as any of its organisation remained, and then, refusing to take office under the Manchus, retired to a mountain near Hêngyang where he spent the rest of his life in study and writing. He seems to have met one or other of the Jesuits, but in spite of this his works show little Western influence. Philosophically Wang Fu-Chih was a materialist and sceptic, strongly opposing the ideas of Wang Yang-Ming and his idealist colleagues, as well as setting his face firmly against any form of superstition. He attacked astrology and phenomenalism, and though supporting Neo-Confucianism, at least in principle, he was not happy about the theory of world-catastrophes; indeed Wang Fu-Chih refused to consider this or other cosmological issues because he thought they lay too far beyond what could be observed or even usefully discussed. In a sense, then, he returned to the more ancient Confucian position, though at a more sophisticated level, so to speak. His work is contained mostly in commentaries on the *I Ching* and in a few smaller books, especially the *Ssu Chieh* (Wait and Analyse) of about 1670, and the *Ssu Wên Lu* (Record of Thoughts and Questionings).

For Wang Fu-Chih reality consisted of matter in continuous motion, and he emphasised the materialist interpretation of Chu Hsi's philosophy, giving *Li* and *chhi* equal importance. 'Apart from phenomena', he wrote, 'there is no Tao.' But his most interesting contribution to Chinese scientific thought was his concern with what the biologist today would call dynamic equilibrium – the control of a complex organism by feedback mechanisms that tend at all times to keep every part of the system constant and in harmony with every other part, whatever is happening in its environment. Forms, according to Wang Fu-Chih, remain recognisably the same for certain periods of time, even though their material composition is in process of continual change, as, for example, a fountain or a flame. Since he unhesitatingly applied this view to life-forms, he may truly be said to have appreciated the existence of physical and chemical changes in a living body – of metabolism, in fact.

Of course he used no such late nineteenth-century term or its equivalent, but the concept was implicit in what he wrote.

To Wang Fu-Chih, the Five Elements were simply a basis for all the different kinds of substance to be found in the natural world, but the myriad forms of things did not, he thought, possess any 'unchanging material substratum'; on the contrary, their material was in constant change as long as they existed. As far as coming into being and passing away were concerned, he believed that things could disperse and return to the Great Undifferentiatedness, to the Origin of the Generative Force of Nature, and that nothing is completely extinguished. 'Life is not creation from nothing, and death is not complete dispersion and destruction.' These ideas of what we may term an 'assembly' of parts and a later return of those parts 'to store', though probably derived from ideas of aggregation and dispersion that went back to the fourth century B.C., acquired in the seventeenth century such precision in the hands of Wang Fu-Chih and men like him, that they virtually amounted to an appreciation of the law of the conservation of matter. Had Wang Fu-Chih ever known of the Western form of this principle, he would surely have recognised it as his own thought.

In spite of his notable work for naturalism and materialism, most of Wang Fu-Chih's time was spent on historical questions. Yet even here his materialism showed itself, allied with the burning patriotism of a man opposed to the foreign domination of the Manchus, a man who had supported the cause of the Ming to the very end, and who wished his epitaph to be only 'The last of the servants of the Ming'. His actual historical work was concerned with differentiating between ancient feudalism and the feudal bureaucracy, and elaborating a theory of social evolution. He praised national heroes, denounced traitors, and dissected the failings of a bureaucratic society: this last occupied a great deal of his attention. It ranged from an attack on corruption in official bureaucracy to a recognition of the importance which a merchant class could have, a class the development of which, he maintained, had been held back by the bureaucracy in spite of its value for the country. In general Wang Fu-Chih adopted some very modern economic attitudes, and it is perhaps little wonder that many Chinese marxists look on him as a forerunner of Marx and Engels.

The rediscovery of Han thought

Two contemporaries of Wang Fu-Chih in the materialist movement were the scholars Yen Yuan (1635–1704) and Li Kung (1659–1733), and they were important because they founded a group which became known as the 'Yen–Li School' or the 'Han Hsüeh Phai' (Back to the

Han Movement). The group attacked the Neo-Confucians with the aim of trying to get back to the ideas of Han scholars some eighteen hundred years earlier, and based their case on a demonstration that Neo-Confucianism was deeply impregnated with Buddhist and Taoist elements. But their greatest significance lies in the fact that it was their work which paved the way for a new approach by Wang Fu-Chih's great eighteenth-century successor Tai Chen (1724–77). Tai Chen had become interested in science early on, and when only twenty wrote a short book entitled *Khao Kung Chi Thu Chu* (Commentary on the Artificers' Record). Later in life he was very active in the recovery of old works on mathematics, wrote a book on the calculating rods and served as one of the learned compilers of the Imperial Manuscript Library. According to his follower Ling Thing-Khan (1757–1809) Tai Chen had a great interest not only in science but in technology as well. He was also the greatest of the few philosophical thinkers that the Chhing dynasty produced.

Starting from the basis of Buddhist and Taoist influences in Neo-Confucianism, Tai Chen set about eradicating these and then constructing his own form of outright materialism. He boldly thrust out the concept of *Li* as a heaven-sent entity lodged in the mind, and assumed that *chhi* alone is able to account for everything, including the highest manifestations of human nature. Tai Chen thus returned to the old idea of the Tao as the Order of Nature, as exhibited in the phenomena explained by the Yin and Yang forces and the Five Elements. He also laid great emphasis on the rediscovery of the meaning of 'pattern' in the term *Li*, a meaning that had gradually become lost due to much poetical and moral paraphrasing, and he thus restored it to its position of 'immanence' in matter-energy, *chhi*. What is more, Tai Chen attacked the idea which had grown up of identifying the pattern principle in Neo-Confucianism with a kind of universal law and of using this as an excuse for justifying the activities and laws of the government of the day.

Meditation and introspection were no way to find out the principles of things, Tai Chen maintained; these can be shown only by 'wide learning, careful investigation, exact thinking, clear reasoning and sincere conduct', not by any flash of sudden 'enlightenment'. Reasoning, he went on to explain, is not something imposed by Heaven on man's physical nature; it is exemplified in every aspect of his being, even in the so-called baser emotions. Here Tai Chen was very modern in outlook, and in direct opposition to a popularly accepted form of Neo-Confucianism so permeated with Buddhism that it had begun to teach that all men's natural desires were essentially evil and should be suppressed. For Tai Chen, on the other hand, the ideal society was one in which these desires and feelings could be freely expressed without injury to others. He insisted that even the great qualities of fellow-feeling, righteousness, decorum and wisdom are simply extensions of the fundamental instincts of nutrition and sex, or the natural urge to preserve life and postpone death, and are not to be sought apart from these urges. Virtue will, therefore, not be the absence or suppression of these desires, but their orderly expression and fulfilment.

Tai Chen also held that the social consequences of regarding *Li* as a heaven-sent principle of illumination in human nature worked great harm in Chinese society. Certainly one could say that the recognition that *Li* was present even in the humblest individual enhanced his dignity and gave him a higher law to which he could appeal when subjected to injustice, but, he pointed out, it was not so helpful when it was appealed to as an authority for the personal judgments of a magistrate. No man's private opinion, Tai Chen urged, should be called *Li*; a view that shows he thoroughly understood how scientific proof is public, not private; something that must be generally accepted, not privately vouchsafed.

Among the dissatisfactions which the Yen–Li School had felt with Neo-Confucianism was its predominantly bookish character, and Yen Yuan, in rediscovering the ancients, found abundant evidence for thinking that their educational methods had been far more practical. Accordingly, when he was asked, in 1694, after a period of studying and practising medicine, to take charge of a new school, he brought about what could have become a revolution in Chinese education by introducing technical and practical subjects. He saw that the school, the Chang Nan Shu Yuan, was provided with a gymnasium, that there were halls filled with machines of war that could be used for practice as well as demonstration, separate rooms for mathematics and geography, an astronomical observatory, and facilities for learning hydraulic engineering, architecture, agriculture, applied chemistry and pyrotechnics. Unfortunately the whole school was destroyed by a severe flood after only a few years, and Yen Yuan died before it could be reorganised and rebuilt. Although it may be that Yen Yuan's enterprise owed something to the Jesuits, it must be remembered that a school of this kind would have been very advanced indeed for Europe, even in the late seventeenth century. But it was not the first Chinese attempt to orientate education towards practical affairs; in Sung times Wang An-Shih had introduced papers on hydraulic engineering, medicine, botany and geography into the Imperial examination system, but that was one of the reforms that did not outlast him.

We have now reached the time when the full weight of the Jesuit introduction of post-Renaissance Western science was making itself felt. Yet the interesting thing is that in spite of the theological standpoint of those who brought this new knowledge, the Chinese tradition of naturalism was still so strong that it could raise up thinkers like Tai Chen, a man whose world-outlook was really more in line with the science of today than it was with that of the Jesuits. During the first thousand years and more of our era the flow of inventions and techniques had been mainly from the East to the West, but during the seventeenth and eighteenth centuries, the reverse was taking place, and Tai Chen took advantage of this. It was he, we are told, who, among other things, widely recommended the Archimedean screw for raising water; this was significant, for the principle of the screw had been unknown in Chinese technology, and when the 'occidental dragon-tail water-raising machine' appeared the whole subject required a clear explanation; this Tai Chen provided in his *Lo Tsu Chhê Chi* (Record of the Class of Helical Machines).

The ' new, or experimental ' philosophy

We are now discussing a period which lies just outside the plan of this book, but it is as well to gain a glimpse of the scientific and technological atmosphere after the arrival of the Jesuits – an atmosphere that was highly productive. For instance, in the first half of the seventeenth century Chiangsu province produced two outstanding optical practitioners, Po Yü and Sun Yün-Chhiu, very reminiscent of that 'higher artisanate' which was so important in the nascent modern science of Europe. Both men constructed telescopes and many other types of optical apparatus, including simple microscopes, magic lanterns, searchlights, etc., and Po Yü may have been one of the several independent inventors of some form of telescope along with the Dutchman Lippershey, the Englishman Leonard Digges, the Italian della Porta, and possibly others. Where he does seem to have had certain priority was in the use of the telescope for gunnery in 1635.

Again, in 1638 Tai Jung wrote the *Chhi Chhi Mu Lüeh* (Enumeration of Strange Machines), concerned with the remarkable machines constructed by his friend Huang Lü-Chuang. Huang made and described – or just described – barometers, thermometers, a humidity meter with dial pointers, siphons, mirrors, microscopes and magnifying glasses, various automata, a kind of moving-picture machine, a device that seems to have been either a crank or pedal cart or a bicycle partly worked by springs and capable of 80 *li* (some 43 kilometres) in one day, an automatic fan, water-piping, improved water-raising machinery and so on.

Tai Jung relates that:

'At Kuang-ling in Chiangsu, Huang Lü-Chuang and I lived some time together. We learnt "occidental" geometry, trigonometry and mechanics, and his ingenuity was greatly improved thereby.

'Huang Lü-Chuang made many very ingenious machines and was never exhausted. Some people were astonished at such strange things, and thought that he must have some magical books or teachers. But I lived all the time with him, and used to joke familiarly with him, and I never saw any such books, and I know that he had no such teacher. He used to say, "What is so strange about these things? Heaven and Earth and all creatures are strange things. Moving like the sky, stationary like the Earth, intelligent like men, leaving traces behind as all things do – how could any natural thing not be considered strange? But none of these things are strange on their own account; there must be some source which is the governor and master – extremely strange [in our eyes], yet not strange to itself – just as paintings have to have a painter and buildings an architect. This may be called strangest of all." So I was astonished at the grandeur of his words.'

Are not Aristotle and the ingenious seventeenth-century chemist Robert Boyle both speaking here through the mouth of a seventeenth-century Chinese? Yet Chuang Tzu and Shen Kua seem no less present.

Perhaps, then, this may be allowed to symbolise the conclusion of the story of the development of scientific thought in China, from its earliest beginnings among the hundreds of schools of philosophers until it merges in the seventeenth century with the world-wide unity of modern science.

14

Buddhist thought

In discussing Neo-Confucianism we made several references to Buddhism, the third of the 'three doctrines' of traditional China, and the only non-indigenous one; now at last we must look at it more closely. For no one from outside Western culture, trying to describe the rise and flowering of science and technology in Europe, could avoid devoting a chapter to the relation of Christian ideas to this development. What he would say we can only speculate: he might, for instance, refer to the importance of the idea of a personal creator, or the reality of a time process (since the incarnation occurred at a particular point in time), or to ideas of a 'democratic' character which gave a value to every soul, and hence perhaps to the observations of Nature made by every person. Here, in discussing Chinese science and technology, we are faced with a similar problem: to decide what effects Buddhism had on science after it was introduced into China. Even if these were mainly negative, we still have to consider factors that inhibited development just as much as those that encouraged it, so we cannot dismiss Buddhism out of hand.

General characteristics
The study of Buddhism is apt to be unsatisfying to both scientists and sinologists because there has been some lack of agreement about what the primitive doctrine really was. The dating of important texts, as with all Indian documents, still causes much uncertainty concerning the general history of the ideas, there was no clearly defined orthodoxy, and there have been many varied and often conflicting opinions about how Mahāyāna Buddhism (the form prevailing in China) differed from the earlier Theravadin (or Hīnayāna) Buddhism. At one extreme it has been claimed that original Buddhism was solely magic and the performance of miracles coupled with hypnotism; at another that Buddha was a follower of a previous philosophical system. All that is

Table 17. *Chronology of the rise of Buddhism*

B.C.	
563–483	Life of the founder of the religion, Gautama Siddhārtha, prince of a small country in northern India, Kapilavasthu. (But some authorities place it a century later.)
483	First Council at Rājagaha.
338	Second Council at Vesālī.
321	Maurya Empire founded by Candragupta (cf. the unification of China by Chhin Shih Huang Ti some ninety years later).
269–237	Reign of Aśoka. This is the earliest time from which any epigraphic evidence bearing on Buddhism exists.
247	Third Council at Pātaliputra.
246	Mission of Mahindra to Ceylon.
2nd cent.	Beginnings of Mahāyāna doctrines, continued under the Kushāna kings in the 1st century B.C.
A.D.	
65	The first date at which we can place the appearance of Buddhist monks and laymen in China. They formed a community at Phêng-chhêng (modern Hsüchow in Chiangsu province) under the protection of a Han prince, Liu Ying, who was also a patron of Taoism. A letter to him from the emperor mentions them. It has been shown that the story of the sending out of ambassadors by the emperor Han Ming Ti (58–75), as the result of a dream, and their subsequent return with books, images, and Buddhist monks in person, is nothing but a pious legend fabricated at the beginning of the 3rd century.
78	Accession of Kaniṣka.
100	Council of Sarvāstivādins under Kaniṣka.
2nd cent.	Rise of the dialectical Mādhyamika School of Nāgārjuna.
148	Arrival of the Parthian Buddhist An-Chhing. Among other missionaries of the late 2nd century Chu Shuo-Fo the Indian, and Chih-Chhan the Yüeh-chih, may be remembered. From this time onward, a vast work of translation of texts went on.
5th cent.	Rise of the idealist Yogācāra School of Vasubandhu and Asaṅga.
6th cent.	Rise of Dignāga's School of Logic (Chhen-Na).
7th cent.	Śāntideva, Dharmakīrti, and the rise of the Tantric Schools.

certain is that the Buddhist philosophical system did come into existence, a system with theories about the nature of the individual, his
career according to the laws of causation, and a doctrine of his final
destiny. Then later, with the Mahāyāna movement, all these problems
were transformed through a new theory of the Enlightened One, a theory
that made him but one of a succession of saviours.

Because of such difficulties it may at the outset be helpful to summarise
the principal dates involved in a table (Table 17), for immediately we
do this, one striking fact appears: the earliest written traditions which
have come down to us do not date from before the fourth century B.C.,
when the *Dīpavamsa* (History of the Island of Sri Lanka) was written,
and it was then another hundred years before the *Mahāvasma* (The
Great Chronicle) appeared. Thus there is nothing in any way equivalent
to the Chinese *Chhun Chhiu* (Spring and Autumn Annals), or even to
the *Shih Chi* (Historical Record): we are therefore more accurately
informed about the life and times of Confucius than about the beginnings
of Buddhism, though these were a century later.

Another important fact is that the Buddhists split into two sects long
before there were any written records at all. However, all sects were
united on certain fundamental points, of which perhaps the most
central was the doctrine of *karma* or retributive destiny. As an explanation of good and evil in this life in terms of good and evil in past lives,
and involving the idea that at death the soul transmigrated to another
body, it was pre-Buddhist; Buddha's insight was to regard the misery
and happiness thus experienced in terms of ethics and morality of past
lives only, not in the performance of ritual or sacrificial acts. The Jains
and other religious men in India believed they could reduce or improve
an individual *karma* by ascetic practices, often carried to extremes, but
all the legends of Buddha's life are agreed that he decisively rejected
this view. His doctrine of *karma* embodied only the 'Four Noble
Truths' – (1) suffering exists; (2) its cause is thirst, craving, or desire;
(3) an overcoming of suffering is possible; and (4) this can be achieved
by the cessation of desire and by self-training along the 'Noble Eightfold
Path'. This included all kinds of psychological exercises as well as
mortifications stopping short however of the more extreme forms of
asceticism.

The notion of retribution was often stated as a kind of doctrine of
cause and effect. The conviction of inevitable consequences contained
in the idea of *karma* was always expressed in terms of cosmic law. This
was connected with a sort of vicious circle from which human beings
had to escape – the *Twelve Nidānas*. These formed a 'cycle' of propositions, since not only did one give rise to another, but also the last

Table 18. *The cycle of the Twelve* Nidānas

IGNORANCE (*avidyā; wu ming* 無明)	causes the appearance of	the AGGREGATES (*saṃskāra; hsing,* 行) (these are considered to mean manifestations of the will)
the AGGREGATES (*saṃskāra*)	cause the appearance of	CONSCIOUSNESS (*vijñāna; shih,* 識)
CONSCIOUSNESS (*vijñāna*)	causes the appearance of	MIND AND BODY (*nāmarūpa; ming sê,* 名色)
MIND AND BODY (*nāmarūpa*) (lit. name and form)	cause the appearance of	the SIX SENSE-ORGANS (*saḍāyatana; liu ju,* 六入)
the SIX SENSE-ORGANS (*saḍāyatana*)	cause the appearance of	CONTACT (*sparśa; chhu,* 觸)
CONTACT (*sparśa*)	causes the appearance of	SENSATION (*vedanā; shou,* 受)
SENSATION (*vedanā*)	causes the appearance of	CRAVING (*tṛṣṇā; ai,* 愛)
CRAVING (*tṛṣṇā*)	causes the appearance of	GRASPING (*upādāna; chhü,* 取)
GRASPING (*upādāna*)	causes the appearance of	COMING INTO EXISTENCE (*bhava; yu,* 有)
COMING INTO EXISTENCE (*bhava*)	causes the appearance of	BIRTH (*jāti; shêng,* 生)
BIRTH (*jāti*)	causes the appearance of	OLD AGE, SICKNESS, DEATH AND ALL MISERIES (*jarāmaraṇa; lao ssu,* 老死)
OLD AGE, ETC., AND ALL MISERIES (*jarāmaraṇa*)	cause the appearance of	IGNORANCE (*avidyā; wu ming,* 無明)

led back again to the first: an endless sequence that was echoed in all Buddhist pictures and preaching by the symbol of the Wheel. Hence the expression 'turning the Wheel of the Law'. It was the aim of Buddhism to set men free from this vicious circle: the Theravadin doctrine emphasising the salvation of the individual by his own efforts, the Mahāyāna the individual's actions in effecting the salvation of others, but in both cases liberation from the world was the central theme. This was to enter the nothingness of *nirvana*.

It can be claimed that the main part of the Twelve Nidānas shows a certain appreciation of sense perceptions and aspects of the activity of the central nervous system, even though the beginning and end of the cycle do not really follow on logically from one another, and demand the eye of religious faith for their acceptance. The Buddhists analysed the body, mind and soul into five 'bundles' of elements that were attached together at birth and scattered at death. Four of these were 'immaterial' – the 'aggregates' of manifestations of the will; consciousness, contact, and, finally, sensation – and one was material, the body (in the sense of physical attraction and bodily pride). In this scheme four elements were incorporated: earth (solidity); water (fluidity); fire (heat); and wind (the quality of motion). These were not the same as the Chinese elements, and it does not seem that they had any notable effect on Chinese non-Buddhist thought.

In the pre-Buddhist times of the sacred Hindu writings of the *Vedas* and the *Upanishads* there was a simple belief in the existence of an individual soul, and then, later, the union of the soul with the universe or with God. The Buddhists, however, strenuously denied the existence of an individual soul, although they did maintain that the constituents that make up an individual continued in subsequent incarnations until they were finally scattered if and when a person reached the status of an *arhat*, a perfected being. They also attacked the opposite, the materialist theory that the individual was annihilated at death. All the same, some Buddhists did introduce the idea of individuality as such, and although this never became orthodox doctrine, the idea was accepted that when an individual migrated, he or she was loaded with the *karma* of his past actions; the *gandharva* (the being to be reborn) was spoken of as entering into the embryo or womb (*garbha*), a view that eventually stimulated the Buddhists to take an interest in embryology. As far as the results of rebirth were concerned, these involved a possible range of careers depending on the merit carried over from a past life and on the evil to be expiated: a man could be reborn as a god, another man, a hungry ghost, an animal, or even as an inmate of one of many hells.

To round off the picture of Buddhism before it entered China, we must remember that in its original form it was a doctrine intended for mildly ascetic hermit monks, living together in a community; it was for them alone that the rules of discipline were formulated. Only later was the religion extended to people in everyday life. Perhaps this is why Buddhism disappeared from India rather readily: once the educated monks and their communities had gone, there was really nothing to distinguish the Buddhist layman from the Hindu, for although the social differentiations of the Hindu caste system had been repudiated by Buddhism, they had never been condemned and fought by the laity. In the monasteries, the Buddhists took over yoga practices like meditation and self-hypnosis as a means to deep insight, while it was

generally believed that these could also confer supernatural powers –
materialisations in various places, levitation, telepathy, invisibility,
and so on – as well as help in the conscious regulation of normally
involuntary bodily functions. Yet all this was the specialised province
of the monk, not the layman. And as far as other gods were concerned,
Buddhism tolerated their worship but never considered that they were
able either to bestow happiness or to act as a basis for morality. Later
on, though, Buddhism found it wise to incorporate all the former gods
of the territories which it conquered, enrolling them as protectors of
the faith on a grand scale, sometimes so successfully, as in Tibet, that
the original doctrine was to a large extent obscured.

Yet whatever befell Buddhism, it never lost its basic attitude of
refusing to answer any questions which concerned matters that were
considered unknowable, questions such as whether or not the universe
is eternal, whether or not it is finite, whether or not the vital principle
is the same as the body, and whether a Buddha does or does not exist
after death. Perhaps this refusal to speculate was another feature that
made Buddhism inimical to carrying out scientific research.

The greater and lesser careers

We can now glance at the two forms into which Buddhism crys-
tallised, the Theravadin and the Mahāyāna, the first primitive and
austere (the 'way of the monks'), the second elaborately liturgical and
concerned with redemption through Buddhist saviours. In essence the
first of these promoted either the development of the individual to
oneness with the universe – what its opponents described as the 'disciple's career' – or else it envisaged the individual's goal as the attainment of the status of a 'solitary enlightened one', a Buddha who does
not preach. This became known as the 'Little Career' (*hināyāna*). On
the other hand, the Mahāyāna fathers stressed the desirability of full
Buddhahood and rejected the 'selfish' ideals of the Hīnayāna: salvation
was now to be the goal of monks and laity alike. Buddha, it was
claimed, had worked for the salvation of everyone, so his followers
should also work to this end, if necessary by a further series of rebirths
and a postponement of the entry into *nirvana*. This was the 'Great
Career'. Of the origin of the Mahāyāna ideas we know little; all that
is certain is that they were already highly developed by the second
century A.D., just in time to be conveyed to China. It was the most
'modern' and attractive presentation of Buddhism, with a world full of
bodhisattvas – spiritual beings and their reincarnations who undertook
to save the world – and who should be worshipped as much as the
historic Buddha himself. *Nirvana* was played down, and even the idea

of oneness with the universe, perhaps with the sound idea that self-culture alone could never lead to salvation of the self; only the salvation of others could achieve this. In taking this last attitude, Buddhism presented a ready-made paradox quite to the taste of the Taoists, and one that they were later to find no difficulty in appropriating.

The great document of this new system was a book written in India in the second century, and translated into Chinese in the fifth as the *Miao Fa Lien Hua Ching* (The Lotus of the Wonderful Law). In this the *bodhisattvas* are praised, promises made of Buddhahood for all, and a great deal said about the cyclic appearance of world catastrophes. Yet later, most laity came to know only of the *bodhisattva* worship, the Four Noble Truths being primarily a matter for monks. However, although for a time this new system could be looked on as an expansion of the old, that became impossible when a new doctrine, the teaching of the Void, the belief in the total unreality of the world of experience, arose. Whereas it had once been possible to consider the cycle of rebirths as occurring in a real universe, now everything was pictured as a delusive shadow play, and release into *nirvana* as release from the need to watch it. This was the teaching of the Mādhyamika School, which probably started about the first century B.C. but was not systematised until the second century A.D. by its greatest figure Nāgārjuna or Lung-Shu, who lived just after its introduction into China.

Buddhist teaching now maintained that everything was in perpetual change, not the same for one instant, with the result that nothing was permanent, nothing real. Shapes, feelings, aggregates that sum up an individual's mental and physical existence, were only illusions. This was a totally negative but very powerful view, so powerful that it is almost impossible to overestimate the importance which its theme of illusion had in Chinese Buddhism. It was this, most of all, that made Buddhism irreconcilable with Taoism and Confucianism and tragically played a part in strangling the development of Chinese science.

The Buddhist evangelisation of China

The first Buddhists arrived in China about the middle of the first century A.D., and from the middle of the next century onwards Buddhist texts poured into the country in an unceasing stream, reaching a peak three hundred years later. Yet as early as the end of the second century a small but significant book, the *Li Huo* (Resolution of Doubts), had appeared. Written by a layman whose family name was Mou and who had lived for some time in Indo-China, where he became acquainted with Buddhism, it contains little relevant to science, yet it was important because it was extremely polite to Confucianism and Taoism, seeking to justify Buddhism by reference to the Chinese classics. Meanwhile many Indian monks were crossing into China, where they spent their time translating imported Buddhist texts into Chinese. It was, of course, impossible for the Chinese to recognise any chronological sequence in these writings, which included both Hīnayāna and Mahāyāna texts, and as a result purely artificial classifications were set up. Thus the theory arose that there had been five periods in Gautama Buddha's preaching. Another characteristic product of Chinese Buddhism was the *Chhan* (Zen) School, a mysticism of the purest quality supposed to have been founded by the Indian Bodhidharma (Ta-Mo), who died about 475. Rejecting all philosophy, it relied exclusively on mystical insight, with intense and prolonged contemplation. *Chhan* has often been considered a type of Buddhism largely due to Taoist influence. Of immense cultural and artistic influence, it was, however, yet another factor unfavourable to a scientific view of the world.

The last type of Chinese Buddhism that needs to be mentioned is the so-called Pure Land (*Chhing Thu*) sect, which believed that by devotion to one of the *bodhisattvas*, Amida, the individual could be reborn in a happy land somewhere in the Far West, where they could listen to particularly efficacious preaching about *nirvana*, though ultimately that concept dropped out and only the pure land remained. This school flowered only in China and Japan, where even now many Buddhists still pray for entry into the pure land. Like *Chhan*, the *Chhing Thu* School perhaps owed more than a little to Taoist ideas of paradise.

Besides these sects, some other philosophical schools grew up, notably the School of Matter as Such (*Chi Sê Tsung*) which distinguished between the fine matter of intangible things and the coarse matter of the everyday world. This was connected with the introduction of Indian ideas on atomism into China. Buddhist thought was always partial to physical speculations, but their effects on Chinese thought were relatively minor. The basic difference between Chinese thought and the Buddhist approach is clear, and was succinctly expressed in the fifth century by Hsiao Tzu-Hsien: 'For Confucius and for Lao Tzu the regulating of the things in this world is the main objective, but for the Buddhists the object is to escape from this world.'

The reaction of Chinese naturalism

One of the chief causes of strain between Buddhism and the indigenous philosophies of China was that however much they tried to escape from the idea of an individual soul, the Buddhists were forced to admit the existence of *something* individual which persisted through successive reincarnations. It was here that they collided head on with the sceptical tradition of the Confucians and the selflessness of the Taoists. Again, the Buddhist doctrine that the visible world was an illusion was a direct contradiction of all Confucian and Taoist teaching, and it is not surprising, therefore, that direct attacks were launched on the Buddhist outlook, and that a Chinese alternative to Buddhism developed. This was Neo-Confucianism, the union of Confucianism with Taoist philosophy, but it did not arise until Sung times for not until then had the points at issue really become clear. The Neo-Confucians, indeed, formed the main opposition to Buddhism, since the Taoists said little, partly because they were extremely busy absorbing Buddhist usages and liturgy into their own rather artificially organised religion, and partly because they had experienced a second flowering of ideas in alchemy and other practical studies which were little affected by Buddhism.

The deep antipathy between Buddhist and Chinese philosophers was well expressed by Hu Yin, who, in the twelfth century, said that ice and glowing coals would mix better than Confucianism and Buddhism. But the greatest of the Neo-Confucian philosophers, who was for ever waging war against the Buddhists, was Chu Hsi, whose important views of the universe we discussed in a previous chapter. Essentially, Chu Hsi emphasised an organic cosmic pattern and rejected the Buddhist doctrine that the natural world was an illusion:

> 'Liao Tzu-Hui wrote as follows to Chu Hsi: "There is only one *Li* of Heaven and Man. The root and the fruit are identical. When the Tao of man['s nature] is perfected, the Tao of Heaven is also perfected. But the realisation of the fruit does not mean separation from its root. Even those whom we regard as sages spoke only of perfecting [the relationships of human life].
>
> 'Now the Buddhists discard man, and discourse [only] on Heaven, thus separating the fruit from the root, as if [they were two horns of a dilemma of which] you must choose one and reject the other...
>
> 'But there are not two *Li* in the universe. How then can the Buddhists take Heaven and man, the root and the fruit, summarily asserting the one and denying the other, and call this a Tao? When their perceptions are so small and incomplete and partial, what possibility is there of the familiar doctrine of a perfect union between the transcendental and the lowly?"'

This last question echoes the immortal words of another Neo-Confucian, Chhêng Hao (d. 1085), who, with respect to the Buddhists, wrote:

'When they strive only to "understand the high" without "studying the low", how can their understanding of the high be right?'

Strong well-reasoned arguments were also raised by Chu Hsi's pupil, Chhen Shun, who did not like the other-worldly idealism of the Buddhists either, since he rightly felt they were inimical to any scientific view of Nature. The Buddhist aim of release from existence was not the proper way for Man to react to Nature, and reincarnation, Chhen Shun argued, left no place for its infinite creativity and novelty. In brief, like his master, Chu Hsi, his Neo-Confucian opposition to Buddhism was essentially that of a scientific view of the world combating a world-denying ascetic faith.

Influences of Buddhism on Chinese science and scientific thought

In spite of their doctrine of the world as an illusion, there are certain specific theories associated with Buddhism which probably had a broadening effect on Chinese minds, and might even have disposed them towards science. One was the conviction of the infinity of space and time, and of the existence of other worlds besides the Earth; another was the notion of recurrent world catastrophes. In these, the Buddhists believed, the sea and the land were turned upside down, and all things reduced to a state of chaos before separating out into the normal world again. This does seem to have borne some useful fruit, for it was this view that was probably partly responsible for the recognition in China of the true nature of fossils.

Yet another point at which Buddhism made contact with Chinese scientific thinking was over the question of biological change. The doctrine of reincarnation naturally aroused interest in those remarkable animal transformations which the Chinese had long studied: if, for instance, birds could turn into mussels, as it was generally supposed they could, it seemed less surprising if men could do this, if their load of *karma* were sufficiently heavy, or even into *pretas* (hungry ghosts) if the load were still worse. Yet this was only one effect of the reincarnation doctrine, dealing with the end of life; what of the beginning of the new life? Interest in this question led to much discussion and finally to a heightened interest in embryology, although the other factors in Buddhism that militated against Chinese science prevented anything useful being done. Interestingly enough, the effect of Buddhism here was somewhat parallel to the strong influence exerted on seventeenth- and eighteenth-century European embryology by Christian theological ideas such as the transmission of original sin and the entry of the soul into the unborn child. But to return to China, an example of the Chinese-Buddhist case will show how the matter stood as far as metamorphosis was concerned. In the twelfth-century *Mêng Chai Pi Than* (Essays from the Mêng Hall) by Chêng Ching-Wang we read:

> 'Chuang Chou said, "All things arise from germs [*chi*] and go back to germs." This is also recorded in the *Lieh Tzu* book, which has a more complete statement. When I lived in the mountains and quietly observed the transformations of things, I saw many examples of it. The outstanding ones are that earthworms turn into lilies, and that wheat, when it has rotted, turns into moths. [Mistaken beliefs, as we now know, that arose from careful observations wrongly interpreted.] From the ordinary principles of things we cannot analyse these phenomena. [One would suppose that] whenever such a transformation occurs, there must be some perception which brings about an inclination for it.
>
> 'Now the change from the earthworm into the lily is a change from a thing which possesses perception into a thing which has none. But the change from wheat grains to moths is just the opposite. . .
>
> 'According to the Buddhists, these changes are brought about by extremely real and pure intentions. From these general and specific causes such phenomena arise. Take the everyday fact of the hen hatching the egg, for example; we know that the egg comes from the hen itself, but how can you explain the fact that a hen can hatch a duck's egg, and even as Chuang Tzu records, that a hen can hatch a swan's egg?'

A biological element can also be found in Buddhist pictures and carvings, as for instance in the Temple of the Sleeping Buddha at Chiu-Chhüan in Kansu province, where there is a statue (Fig. 29) showing a monk undergoing a spiritual metamorphosis, moulting off the 'old man' by pulling his former skin apart with his hands.

But behind all these points of tentative contact between Buddhist ideas and the developing interest in the sciences of Nature is the fact that Buddhism introduced into China a wealth of highly sophisticated discussions about logic and the theory and nature of knowledge. Buddhism, and other Indian theoretical systems that accompanied it, were often at least as subtle as any of the great philosophies of Europe, and some, not least those concerned with dialectical logic, reinforced similar

Fig. 29. Metamorphosis in Buddhist iconography. One of the statues in the temple of the Sleeping Buddha at

thinking already present in Taoist and Mohist schools. Unfortunately, though, there seems no evidence that these highly developed modes of thought had any great effect on scientific speculation or study in China.

Tantrism and its relation with Taoism

There is, however, one other aspect of the relationship of Buddhism to science that must be mentioned, for there was another side to Buddhism, strangely different from the ascetic practice and philosophical idealism so far mentioned. This was an almost Taoist department of Buddhism known as Tantrism.

The Tantras were late sacred texts on the borderline between Hinduism and Buddhism, produced in India at some time certainly not earlier than the sixth century A.D. The practices accompanying them were sometimes open to all, sometimes confined to the select few, and at first sight very odd indeed. Worship of personal gods was prominent, but more characteristic was a strongly magical element which contained 'words of power', talismans, amulets, hand gestures and other charms, all of which were mentioned in the texts. Tantrists also adopted in their symbolism what we might call 'electrical imagery', but known to them as the 'thunderbolt'. One can see at once that here was a system of thought closely akin to the shamanist and magical side of Taoism and, on the principle that magic and science were once united in a single complex of manual operations, in Tantrism, if anywhere, Buddhism might have produced some contribution to science.

Just as early and mediaeval Taoism was always interested in sex, so was Tantrism. The thunderbolt, for example, was identified with the male external generative organ, while the lotus – so characteristic of Buddhist picture symbolism – was identified with the female. Essentially, the theological doctrine was that the mystical or divine energy of a god (or of a Buddha) resided in his female counterpart or *shakti*, from whom he received it in an eternal embrace. The logical conclusion that followed from this was that the earthly *yogi*, seeking for perfection, must also embrace his *yogini* in a sexual union prepared for and conducted with special rites and ceremonies. There also followed the worship of women as a preliminary to sexual union, and the whole formed, indeed, a quite remarkable parallel to the early practices of Taoism. We must not, of course, judge these ideas by the canons of two thousand years of Gnostic and Manichaean anti-sexuality transmitted through many of the Christian fathers, but rather consider them dispassionately as the attitudes of a different culture from which we might have something to learn. If we do this, we cannot but concede that the whole perspective may well have been associated with a magico-scientific outlook, for apart from anything else sexual symbolism played an enormous part in the language and thought of alchemy. Could it not even be that the very concept of chemical reaction arose out of the congress of the human sexes?

Indian Buddhist Tantrism appears to have come to China in the eighth century A.D., although magic formulae had certainly arrived at least five hundred years earlier. One of the most important Chinese Tantrists, the monk I-Hsing, lived at this time, and the fact that he was the most eminent astronomer, mathematician and designer of clockwork of his day may well indicate the significance of this form of Buddhism for all kinds of observational and experimental science. Actually one cannot but wonder whether Tantrism was really an Indian invention: close inspection makes it seem at least possible that the whole thing was originally Taoist after all. There are many parallels, and when we consider that Taoist sexual theories and practices flourished from the second century to the sixth, this idea does not stretch credibility too far. Indeed it even seems possible that Tantrism developed partly from Taoist ideas exported from China through Assam, and then came back centuries later: if this were so, it would not be an isolated case of foreigners amicably instructing the Chinese in matters with which the Chinese were already familiar. On the other hand, the sexual element in Indian religion had been so important from early times that Buddhist Tantrism may have been a kind of hybrid of Buddhism and Hinduism.

When all is said and done, though, Buddhism does not seem to have helped the development of science in China. It was essentially inimical to it as any philosophy built on a profound rejection of the world was bound to be. Buddhist ideas of law might conceivably have stimulated the notion of Laws of Nature, if only the *karma* doctrine had been less exclusively ethical. But cosmic inevitability was in a way only a vestment with which the Buddhists clothed their profound religious belief in divine justice. Buddhism, it is true, was a great civilising force in Central Asia, but in China, where there was already a civilisation of a high order, matters were rather different. There Buddhism introduced that element of universal compassion that neither Taoism nor Confucianism, rooted as they were in family-ridden Chinese society, could produce. Despairing in its philosophy it might be, and for the scientist perverse, but its later practice was often plainly and recognisably that of universal love.

15

The Legalists

In describing the ideas of Tai Chen a few pages ago we saw how the concept of law was never far away from scientific and social thought. For the rest of this book we shall be concerned with the very notion of law, whether in human society or in Nature. And first we must return to the fourth century B.C. and look at the philosophical school known as the Legalists.

If one is tempted to become impatient with Confucian moralising, one has only to read the writings of the Legalists to come back to Confucianism with open arms, embracing their profound humanitarian resistance to Legalist authoritarianism. Yet the Legalists are important from the point of view of Chinese science because their beliefs raise, in an acute form, the problem of the relation of juridical law and the laws of Nature.

It has often been said that the peculiar glory of Chinese law lay in the fact that throughout its history (after the failure of the Legalists) it remained indissolubly connected with custom, custom that was based on easily demonstrable ethical principles. This is known as natural law, as opposed to laws laid down by princes or parliaments not necessarily connected with morality but which may be useful to everyone. So, for example, the prohibition of the killing of fathers by their sons would be natural law, but the decision that vehicles shall always keep to the left (or the right) sides of roads would be a positive law. In China for the most part the enactments and codifications of positive law were reduced to an absolute minimum. Nevertheless such an attitude makes for an intellectual climate unfavourable to the development of systematised scientific thought, and we must therefore give at least a passing glance at the Legalists and their ideas.

The elaboration of China's first criminal code goes back to the sixth century B.C., but the rise of the Legalists did not occur until two centuries later, and they did not reach their position of real dominance until the third century B.C., when their policies enabled the last prince of Chhin to become the first emperor of a unified China. But, as we have already seen, their harsh measures brought revulsion and led to the milder rule of the Han some twenty years later. The basic idea of the Legalists was simple: they thought that *li* – the complex of customs, usages, ceremonies and compromises – paternalistically administered according to Confucian ideals, was nowhere near forceful enough for a strong government. Their watchword was *fa* (法), positive law, and especially *hsien ting fa* or 'laws fixed beforehand', to which everyone in the state, from the patricians to the lowest sort of slave, was bound to submit subject to sanctions of the severest kind. Here there was a strong democratic element which explains the recent approval of the Legalists in China. But for this end, the lawgiving prince must, they said, surround himself with an aura of authority and majesty.

The central conception of *fa* enacted by the lawgiving prince was to be independent of accepted morality or popular goodwill. The laws were formulated, published in every district, and thus made it clear to all how they must behave. 'Law is the authoritative principle for the people, and the basis of government; it is what shapes the people', said Shang Yang, the Legalist whose reorganisation of the state of Chhin paved the way for the Chhin dynasty. If the law were strong, the country would be strong, and punishments should be deterrent in the highest degree; even the lightest crimes should be severely punished, for where small offences do not occur, great crimes will not follow. It should be made worse for people to fall into the hands of their own police than to fight the forces of an enemy state in battle, and the timorous should be put to death in the manner they most hate. There should also be an elaborate system of reporting and denunciation: even between members of the same family.

The Legalists were conscious of a conflict between their theoretically constructed *fa* on the one hand, and ethics, equity and what one might call commonsense, on the other. Obedience, not virtue, was what the state required; indeed the Legalists drew up a list of 'Six Parasitic Functions' (literally 'Six Lice') – care for old age, living (without employment) on others, beauty, love, ambition, and 'virtuous conduct'. These functions, which were later extended in number, all sapped the power of an authoritarian state. There was also a list of the 'Five Gnawing Worms' which destroy the state: the Confucian scholar praising the sage-kings, and discussing benevolence and righteousness; the clever talker (a hit at the Logicians?); the soldier of fortune collecting troops of adherents; the merchant and the artisan amassing wealth; and the official thinking only of personal interest. Needless to say, the

Legalists had many arguments with the Confucians, for they believed a son should not conceal a father's crime and that officials should be chosen for their ruthlessness. Naturally they glorified war. The Legalists also set great store by agriculture, since good harvests contributed mightily to the power of the state.

But there was one feature in Legalism of interest for the history of Chinese science, its tendency to the quantitative, expressing things in numerical terms. This accounts for the remarkable standardisations introduced by the first Chhin emperor, the unification of the script, the normalisation of weights and measures, even the gauge of chariot-wheels. There were degrees of penalties; everything was assessed by precise quantities, exact weighings, accurate measurements. There was nothing that did not have to be laid down in full numerical detail and, it is worth noting, 'standard' is the oldest meaning of *fa*. Recently Chinese archaeologists have discovered in Chhin tombs thousands of bamboo and wood manuscripts which bring this out very clearly. Consequently the Legalists had a passion for reducing complex personal relationships to formulae of almost mathematical simplicity, and in the end this made them representatives of a kind of mechanistic materialism. If this had succeeded, they might have constructed an abstract system of human relationships analogous to the great structure of Roman law. They failed because they believed somewhat ingenuously that human emotions and human conduct could be measured quantitatively as could grams of salt or metres of cloth.

The Legalists knew, however, that in their quantisation of everything they were doing something new, and it seems that this attitude was partly brought about by new technological developments, e.g. in metallurgy. These could also have coloured their new social theories, and perhaps it was from the arsenals of Chhin that the impetus for so authoritarian a régime was derived. But this is a question which must be left to a later volume; now we are concerned only with the question whether quantised positive law could have given rise to quantised 'laws' of natural science. That this did not happen was due not only to the political failure of the Legalist system, but also the subsequent weakening of the idea of the all-powerful nature of law laid down by a princely lawgiver. Beyond this there was a lack of any concept of a Creator-God, a celestial lawgiver, one who could pre-ordain 'Laws of Nature', just as an earthly lawgiver enacted a juristic code. Hence there could be no contact between the sociology of the Legalists and the nascent sciences of Nature.

16

Human law and the laws of Nature

One of the most basic background ideas of the modern natural sciences is embodied in a phrase on the lips of everyone who talks about them – the laws of Nature. Instinctively recognised as a metaphor by most people, it does nevertheless express the regularities of behaviour of objects in the world we live in, the invariable tendency of heavy things to fall, the fact that water always flows downhill, or that the sun will rise again tomorrow. In the preceding pages we have found ourselves discussing law already twice, first the moral law of the Buddhists, and then the positive law of the Legalists; and in each case we saw good reasons why their conceptions of law never came to be applied to natural phenomena in China. Let us end this volume by enquiring whether in ancient China there ever arose any concept of 'laws of Nature', any notion of 'law' in the scientific sense to describe the reliable behaviour of the natural world.

Can one find then any phrase in the Chinese philosophical and scientific writings which asks to be translated as 'law of Nature', and in no other way? This is obviously an important question; and one which must be answered if we are to have a true understanding of Chinese science. In the West, in Europe, there has always been a close connection between the idea of natural law and the appreciation of the recurrent operations of Nature. Natural law in the juristic sense, law which it is natural for all men to obey, has always been closely linked in men's minds with the concept of laws enacted by God the creator for Nature; both go back to the same common root. But what of the Chinese? Here the answer is nowhere near so straightforward, nor so easy to find.

In Western civilisation there is no doubt that one of the oldest notions was that just as earthly lawgivers promulgated codes of positive law to be obeyed by men, so also the celestial and supreme rational creator deity had laid down a series of laws which were to be obeyed by minerals, crystals, plants, animals and the stars in their courses. It

276

is hard to say precisely when in European or Islamic history the term 'law of Nature' was first used in a scientific sense, but there is ample evidence to show that by the eighteenth century it was current coin in speech. Most Europeans are acquainted with those Newtonian words of 1796:

> Praise the Lord for he hath spoken,
> Worlds his mighty voice obeyed;
> Laws, which never shall be broken,
> For their guidance he hath made.

Yet it is a fact that such words could never have been written at any time by a Chinese scholar, brought up in the patterns of thought of his own culture, simply because world concepts there were radically different from those in the West. What we would like to find out is why this was so.

To answer the question we must consider four points: first, the basic concepts involved; secondly, the development of Chinese law and jurisprudence; thirdly, a sketch of the history of the differentiation in Europe of the ideas of natural law and of the laws of Nature; and last a comparison between the unfolding of thought about these matters in China and in the West. Of course our aim in doing this will also be to see whether this gives us a further clue to those factors in Chinese culture which inhibited the rise in China of modern science and technology.

The common root of natural law and the laws of Nature

It seems that in primitive society the earliest law was the law of unwritten custom. There were no pre-ordained commands, and the sanctions that grew up were derived from the moral disapproval of the community. Only gradually, as society developed and became differentiated into classes, did a body of judgments develop, and with the growth of a central state, these judgments became broader than those which society had previously followed; in fact lawgivers arose. Laws now became embodied in the statutes of the State, laws which incorporated not only the immemorial customs of tribal culture, but also those which were deemed to be good for the welfare of the community, and these were laws which did not necessarily have an ethical basis. So 'positive' law arose – the law, the commands, of an earthly ruler; obedience was a citizen's obligation, and transgressions were punished by precisely specified penalties.

In Chinese thought all this was expressed by the word *fa*, just as the customs of society based on ethics, ancient taboos and ceremonies and sacrifices were covered by the word *li*. Yet the Chinese do not seem to have had any strong predilection for *fa*, for positive law; after the defeat of the Legalists at the end of the third century B.C., it was reduced to a minimum, and custom returned to its former dominance. Confucian lawyers exalted arbitration, compromise and ancient usage, confining positive law to purely criminal matters.

In the West, in Roman law, two codes were recognised, the civil laws of a specific people or state, positive law or *lex legale*, and the law of nations (*jus gentium*), which was more or less equivalent to natural law (*jus naturale*). The *jus gentium* was presumed to follow *jus naturale* unless anything appeared to the contrary: indeed, in Roman law their identity was assumed, even if not very safely, for some customs would certainly not be self-evident to natural reason, and there were rules (the undesirability of slavery, for instance) which deserved to be recognised by all mankind, but in fact were not. The traditional explanation for the origin of this kind of 'natural' law is that it was necessitated by the increasing number of merchants and other foreigners at Rome, who were not citizens, and who, therefore, were not subject to Roman law. They wished to be judged by their own laws, and the Roman jurisconsults found that the best they could do for such a community was to take a kind of lowest common denominator of the usages of all known peoples, and attempt to codify what appeared to be nearest to justice for the greatest number of people. Thus it was that the concept of 'natural law' originated – or, at least, this is one explanation of it, one that will do sufficiently well for our purpose. Natural law was, then, what men everywhere thought to be naturally right.

The difference between positive (conventional) law and natural law is also to be found in Aristotle. In the *Nicomachean Ethics* he says:

> 'Political justice is of two kinds, one natural and the other conventional. A rule of justice is natural when it has the same validity everywhere, and does not depend on our accepting it or not. A rule of justice is conventional when in the first instance it may be settled in one way or the other indifferently – though once having been settled it is not indifferent; for example that the ransom for a prisoner shall be a *mina* [a gold coin], or that a sacrifice shall consist of a goat and not of two sheep, etc. . . . Some people think that all rules of justice are merely conventional, because whereas [a law of] Nature is immutable and has the same validity everywhere, as fire burns both here and in Persia, rules of justice are seen to vary . . . This is not absolutely true, but only with qualifications . . . But

nevertheless there is such a thing as natural justice as well as justice not ordained by Nature, and it is easy to see which rules of justice, though not absolute, are natural, and which are not natural but legal and conventional, both sorts alike being variable.'

This is an interesting passage because Aristotle not only claims that quantitative and non-ethical matters – the size of a ransom, for instance – can only be settled by positive legislation, but also trembles on the verge of speaking of laws of Nature in the scientific sense. In China, however, we find an immediate contrast: there it was hardly possible to have a *jus gentium* since, owing to the isolation and dominance of Chinese civilisation, there were no other respectable nations from the practices of which a universal international law could be deduced. There was, however, a *jus naturale* which the sage kings and the people had always accepted; this was what the Confucians called *li* 禮.

Natural law and positive law in Chinese jurisprudence

Peoples of Western civilisation have lived under the Graeco-Roman conception of law, a system which has also inspired large sections of Islamic law. In the West law has always been revered as something more or less sacrosanct, imposing itself on everyone, great or small, defining and regulating the conditions of all forms of social activity. But as one passes to the East this picture changes, and in China law and jurisprudence took an inferior place to the powerful body of spiritual and moral values which she created and which were diffused among neighbouring cultures. China recognised natural law and exalted the rules of morality; Chinese positive law was above all administrative. And this is echoed by the fact that, though a nation of scholars, China produced relatively few famous judges, fewer commentators and theoreticians of legal matters, and no celebrated advocates at all.

The oldest notice of a legal code occurs in the Lü Hsing chapter in the *Shu Ching* (Historical Classic): it is of the Chou period though its precise date is uncertain. The oldest datable code is that in the *Tso Chuan* for 535 B.C., and yet even here, at the beginning of the story, there appears that uncompromising objection to codifications which characterised Confucian thought throughout Chinese history:

'In the third month the people of the state of Chêng made [metal cauldrons on which were inscribed laws relating to] the punishment [of crimes]. Shu Hsiang wrote to Tzu-Chhan [i.e. Kungsun Chhiao, prime minister of Chêng], saying:

'"Formerly, Sir, I took you as my model. Now I can no

longer do so. The ancient kings, who weighed matters very carefully before establishing ordinances, did not [write down] their system of punishments, fearing to awaken a litigious spirit among the people. But since all crimes cannot be prevented, they set up the barrier of righteousness, bound the people by administrative ordinances, treated them according to just usage, guarded them with good faith, and surrounded them with benevolence.... But when the people know that there are laws regulating punishments, they have no respectful fear of authority. A litigious spirit awakes, invoking the letter of the law, and trusting that evil actions will not fall under its provisions. Government becomes impossible... Sir, I have heard it said that a state has most laws when it is about to perish."'

Thus, from the start, the supple and personal relations of *li* were felt to be preferable to the rigidity of *fa*. As a later insertion in the *Tso Chuan* put it:

'In the winter Ju-Pin was fortified. The inhabitants of the state of Chin were forced to contribute 220 kilos of iron to make cauldrons on which the penal laws were inscribed. They were those of Fan Hsüan-Tzu. Confucius said, "I fear that Chin is going to destruction. If its government would observe the laws which its founder prince received from his brother, it would direct the people rightly... Now the people will study the laws on the cauldrons and be content with that; they will have no respect for men of high rank."'

This patrician and Confucian concept, enshrined in the proverb 'for each new law a new way of circumventing it will arise', passed unchanged down the centuries. Not until the arrival of modern scientific techniques, such as the measurement of specific gravity or individual differences like finger-prints, was it really possible to control many cases of fraud or abuse in organised society. The Chinese disliked any over-elaboration of laws, yet paradoxically it was they who by the thirteenth century had laid the foundations of forensic medicine (p. 247) and it was they, not Europeans, who discovered finger-printing. Yet may be their proverbial wisdom is wisdom still.

Nothing much is left of the provisions of the earliest Chinese codes, but we do know that the system which proved to be the ancestor of all the later ones was that of Li Khuei, minister of the state of Wei, about 400 B.C. His code was called the *Fa Ching* (Juristic Classic), and though it has long been lost, the headings of its contents have been preserved; they concerned robbery, brigandage, imprisonment, arrest, as well as miscellaneous matters and definitions, and it is these divisions that are to be found in all subsequent codes. Later on, sections were added concerning the census, the family and marriage, vassal labour and military service, and laws connected with the imperial household and the court. This larger corpus formed the Great Code of Shusun Thung and other jurists of the Han dynasty, and although completely lost by the time of the Sui (sixth century A.D.), a good idea of Han practice can still be obtained by reading the *Chhien Han Shu* (History of the Former Han Dynasty) which was written about A.D. 100.

The Chin, Chhi, and other dynasties had their own codes, but little is known of them since, again, all were lost before the Sui. From what information there is, however, they seem to have followed the model of the *Fa Ching* and the Han code in being almost exclusively concerned with criminal matters and government ordinances about taxation. Civil law remained extremely under-developed, but over the centuries the severity of the earliest codes was softened, especially the practices of mutilation and the extermination of a criminal's entire family. Some Confucian scholars wrote about legislation, and there was a period, in the third century A.D., when there were even ten schools of jurists commenting on their teachings, and providing interpretations that passed into common use.

In certain respects the Chinese legal mentality was sometimes ahead of that in Europe. There was a remarkably early appearance of forensic medicine in China, as we have seen, with books on the subject from the Chin (fourth century A.D.) onwards, but given the pattern of Chinese thought this was not an unnatural development since the spirit of *li* insisted that the utmost should be done to prevent guilt being fixed on the innocent. Indeed, such care was taken in this respect that Europeans visiting China some ten centuries later commented admiringly on it. In Europe itself conditions in prisons at this time were often horrific, with torture and flogging prevalent, but what seem to have struck sixteenth-century observers above all about the operation of Chinese law was the system of reviewing all cases in which the death sentence had been pronounced. This is hardly in line with the common belief that human life was always cheaper in China than in Europe, and the opposite testimony of the Portuguese can be independently reinforced from the records of a fifteenth-century Muslim writer. This attitude, this determination to try all possible methods and tests to enable a magistrate to decide a criminal case, fitted in well with the empirical character of the Chinese study of Nature. Even if the standard of scientific enquiry was not sufficiently high to sort out the sound tests from those which rested on a more superstitious basis, the Chinese practice was far more civilised than anything in Europe before the eighteenth century.

The law and phenomenalism. What has been said so far confirms the conclusion suggested already here and in the chapter on the School of Legalists (Chapter 15): in China there was always a struggle between systematic law and the law that the officials administered paternalistically, where every case was considered on its merits and in accordance with *li*. The full force of the meaning behind *li* was profound, and could not be divorced from the customs, usages and ceremonies which epitomised it. The significance of these was deep, lying not merely in the fact that they had arisen because they agreed with the intinctive feeling of rightness experienced by the Chinese, but also in the conviction that they accorded with the 'will of Heaven', with the structure of the whole universe. Hence the basic disquiet aroused in the Chinese mind by crimes, or even by disputes, since these were felt to be disturbances in the Order of Nature. As early as the *Shu Ching* we find evidence of this: for instance, it is stated that excessive rain is a sign of the emperor's injustice, that prolonged drought indicates he is making serious mistakes, while intense heat accuses him of negligence, extreme cold lack of consideration, and strong winds, curiously enough, show that he is being apathetic. Again, in the *Chou Li* (Record of the Rites of Chou), as well as in many other ancient texts, there is the additional idea that punishments can only be carried out in the autumn, when all things are dying: to execute criminals in the spring would have a deleterious effect on growing crops. It is, indeed, as if the Chinese saw phenomena in Heaven and Earth running along parallel strands in time, perturbing events in one strand giving rise to perturbing effects in the other.

We have already come across many examples of this 'phenomenalist' picture when we discussed Wang Chhung and the sceptics (Chapter 11), and the fundamental ideas behind Chinese science (Chapter 10). Wang Chhung, it may be remembered, maintained that excessive seasonal heat and cold did *not* depend on the ruler's joy and anger, that plagues of tigers and grain-eating insects were *not* due to the wickedness of secretaries and minor officials. Yet in spite of this scepticism, the old outlook, embodied in earlier works like the *Tso Chuan*, prevailed. The emperor embodied in himself and, by extension, in his bureaucracy, the system of semi-magical relationships believed to exist between man and the cosmos, relationships which, in primitive times, had been maintained in good order by folk festivals and folk ceremonies.

This phenomenalist outlook was not exclusively Chinese. There was a similar, though less elaborate, conviction among the early Greeks which had obvious connections with theories of the microcosm and the macrocosm. It arose from a kind of projection of the internal relationships of the tribe on to external Nature, and was later to lead men like the Greek physician Galen in the second century A.D. to use the concept of justice in his descriptions of the anatomy of the body. Parts differ in size because Nature had apportioned them according to their usefulness: was this not just? Some parts have fewer nerves than others, but again this is just because they do not need so much sensitivity. Nature, in fact, is always just in all that she does.

We are here in the presence of a thorough parallelism at the three levels of the cosmos, human society and the human body. However, the Western conception was deeply different from the Chinese. The West saw justice and law at all levels closely associated with personalised beings, enacting laws and administering them. The Chinese saw only that righteousness embodied in good custom represented the harmony necessary for the existence and function of the social organism. They recognised, too, the harmony of the heavens, and, if pressed, would have admitted harmony in the individual body as well. But these harmonies were spontaneous, not decreed. Discord in one was echoed by disharmony in the others. In China this phenomenalist conviction gave no stimulus to the idea of the laws of Nature. In the West, on the other hand, the Greek ideas were important elements in the stream of thought which led to the Universal Law of Nature and Life of the Stoic philosophers. In China Wang Chhung's views on phenomenalism were based on the absurdity of supposing that the heavens echoed the petty doings of men, a move against the human-centred universe of phenomenalism, but they did not help in bringing about any conception of scientific law. Such an anthropocentric universe was not rejected in the West until later, but when this did happen, it was possible to break away to a sun-centred universe and still retain the idea of universal law in Nature.

If, as in ancient China, all crimes and disputes were looked on not primarily as ruptures of a purely human legal code, but rather as ominous disturbances in man's connections with Nature, this would pre-suppose so subtle a complex of causal connections that any positive law would seem unsatisfactory. Indeed, the Thang code of the seventh century A.D. specifically suggests that it is dangerous and ominous to 'leave *li* and engage in legally fixed punishments'. With an outlook of this kind there was no place for law in the Latin sense. Rights of individuals were not even guaranteed by law: there were only duties and mutual compromises governed by the ideas of order, responsibility, hierarchy and harmony. The supreme ideal was always to demonstrate a justness, a ritual moderation. To take advantage of one's position, to invoke one's rights, was looked on askance, as it always has been in China. The great art was to give way on certain points and so accumulate an invisible fund of merit whereby one can later obtain advantages in other directions. To this day the Chinese method is to fix responsibility in terms not of 'who has done something' but of 'what has happened'. Only when something has happened and been examined can responsibility be assigned.

Social aspects of law: Chinese and Greek. The full significance of the interplay between *li* and *fa* can, however, only be understood with reference to their relationships to social classes. In feudal times, for instance, it was natural enough that the feudal lords should not consider themselves subject to the positive laws which they promulgated. *Li* was therefore the code of honour among ruling groups, *fa* the ordinances to which the common people were subject. In the *Li Chi* (Record of Rites) of the first century B.C., it is explicitly stated that '*li* does not reach down to the people, *hsing* (punishments or penal statutes) do not reach up to the great officers', a statement that throws further light on the earlier opposition to codification in the sixth century B.C. by Shu Hsiang and Confucius. They were against codification not only because it would lead to litigation and even obstruction on the part of commoners, but also because it embodied the encroachment of fixed laws on the whole class of the feudal nobility. It was precisely this extension, of course, which was carried out by the Legalists, and which paved the way for bureaucratism. And as the Confucians were later the operators of the bureaucratic machine, they in turn became the jurists of positive law. Yet for centuries the fluidity of *li* retained much of its social prestige, and was so much more in keeping with Chinese philosophy than *fa* that it was still predominant even after bureaucratism had become really well established. All this, of course, reveals the meaning of the phrase in the Thang code: 'he who leaves *li* will fall into *hsing*'; in other words if one does not follow the behaviour felt to be ethically right, one will find oneself caught in the net of criminal law. In the event, the flexibility of *li* invariably worked out in favour of the privileged bureaucratic class, and gradations of punishments according to rank among the officials persisted long afterwards into the code of the Chhing (Manchu) dynasty in the nineteenth century.

No doubt *li* and *fa* may be looked on in more than one sociological context. Centralising tendencies and demands for devolution were very delicately balanced in ancient and mediaeval Chinese society: *fa* suited the bureaucratic irregation administrators, *Tao* suited the self-contained rural communities; perhaps *li* was the ultimate compromise between the centre and the periphery of the social organism.

Ancient Greek, as opposed to ancient Roman, law shared to a considerable extent the early Chinese preference for equity and arbitration rather than abstract formulae. The Romans seem to have set their sights on a quest for legal certainty, unrealisable though that might be; the Chinese and the Greeks on the individualisation of cases. The Romans sought what may almost be termed a masculine element, echoing a society in which the power of the father was carried to an extreme; the Chinese preferred a lenience and flexibility that had more of the feminine element about it, a contrast which emphasises the Taoist component in Chinese thought. Indeed, it is probably no exaggeration to say that when Han Confucianism triumphed over the excessive 'maleness' of the Chhin Legalists, it did so partly by accepting from Taoism an attitude to law which rejected the search for a code fixed beforehand, and granted magistrates the widest freedom to follow principles of equity, arbitration and 'natural' law. And this question of the relative rôles of equity and the individualisation of cases on the one hand, and positive law on the other, is still a live issue. In our own world there is the problem of what weight is to be placed on findings in trial courts where the psychology of jurors and defendants, as well as the behaviour of witnesses, can be taken into account, compared with courts of appeal which can only work on rules and interpretations.

Stages in the Mesopotamian–European differentiation of natural law and the laws of Nature

We must turn now to the third part of our argument – the stages of development in Western civilisation of the ideas of natural law and the laws of Nature.

There can be little doubt that the conception of a celestial lawgiver 'legislating' for non-human natural phenomena had its first origins among the Babylonians. By 2000 B.C. the Sun-god Marduk is pictured as a celestial lawgiver who prescribes the laws for the star-gods and 'fixes their bounds': he it is who maintains the stars in their paths by giving 'commands' and 'decrees'. Later we find the pre-Socratic Greek philosophers of the sixth century B.C. speaking much of 'necessity' if not specifically of 'law'. Anaximander (*c.* 560 B.C.) talks of the forces of Nature paying fines and penalties to each other, and Heraclitus (*c.* 500 B.C.) says 'The Sun will not transgress his measures; otherwise the Erinyes, the bailiffs of Dike [the goddess of justice], will find him out.'

Here regularity is already assumed as an obvious empirical fact, and the idea, at least, of law is implied since sanctions are mentioned. Indeed Heraclitus refers to a divine law by which human laws are nourished, a reference that may include non-human Nature as well as human society, since the divine law is 'common to all things', all-powerful, all-sufficing. Nevertheless, the conception of Zeus in the older Greek poets pictures him as giving laws to gods and men, but not to Nature, for he himself was not a creator. The post-Socratic philosopher Demosthenes, however, a contemporary of Mo Ti and Mêng Kho in the fourth century B.C., does use the word 'law' in its most general sense when he says: 'Since also the whole world, and things divine, and what we call the seasons, appear, if we may trust what we see, to be regulated by Law and Order.'

Aristotle never uses the law metaphor although he occasionally comes within an ace of doing so, while Plato adopts it only once. The governance of the whole world by law seems, in fact, to be peculiarly Stoic, no doubt because most of the thinkers of that school, founded about 300 B.C., maintained that Zeus was nothing more than Universal Law. Yet something of this idea may have been implicit in the use by the Pythagoreans, Platonists and the followers of Aristotle of the word 'cosmos' for the universe, since cosmos had connotations with government and order. Nevertheless, strong support for the quite definite conception of the Stoics probably came from Babylonia by way of the astrologers and star-clerks who began to spread throughout the Mediterranean world around 300 B.C. And since Stoic influence was so strong at Rome, it was inevitable that these very broad conceptions should have their effects in the development of the idea of a natural law, common to all men, whatever might be their cultures and local customs. Cicero, in the second century B.C., reflects this when he says: 'The universe obeys God, seas and land obey the universe, and human life is subject to the decrees of the Supreme Law.' Yet, curiously, it is in a poet, Ovid, in the next century, that we find the clearest statements of laws in the non-human world. He does not hesitate to use *lex* (law) for astronomical motions: '*qua sidera lege mearent*' (by what laws the stars move), he writes, when speaking of the teaching of Pythagoras; and elsewhere, complaining of the faithlessness of a friend, he says it is monstrous enough to make the Sun go backward, rivers flow uphill and 'all things proceed reversing Nature's laws'.

The most important scientific school of Greek times was the Epicurean, yet neither Democritus (*c.* 460 B.C.) nor Lucretius (first century B.C.), both of whom powerfully advocated natural and causal explanations, ever spoke of the laws of Nature. There is only one place in the whole of Lucretius' *De Rerum Natura* (On the Nature of Things) where he used the term in its later sense, and this is when he is denying the existence of chimaeras (which combine organs of different origin in the same body as, for instance, in a centaur). These are impossible, Lucretius says, because parts of a body can combine only if they are adapted to each other, and all animals are 'bound by these laws' (*teneris legibus hisce*). This apparently surprising omission was probably due to Epicurean theology, in which the concept of laws of Nature in the strictest sense was not permissible, since the Epicureans taught that the gods had not created the world and took no interest in it; indeed this was most likely the reason why Epicurean philosophers always talked of 'principles', not 'laws'.

Far more certain, however, as another contributory line of thought, was that which emanated from the Hebrews, or was transmitted by them from Babylonia. This was the idea of a body of laws laid down by a transcendent God and covering the actions of both Man and the rest of Nature. It is frequently met with, for the concept of a divine lawgiver was one of the most central themes of Israel. The influence this outlook had on all Western thinking of the Christian era would be hard to overestimate; texts like 'Thou hast set a bound that they [the waters] may not pass over' (Psalm 104: 9) and 'He hath also stablished them for ever and ever: he hath made a decree which they shall not pass' (Psalm 148: 6) were part of everyday thinking. Furthermore, the Jews developed a kind of natural law applying to all men; somewhat analogous to the *jus gentium* of Roman law, it is to be found in the 'Seven Commandments for the Descendants of Noah', which at times was liable to conflict with Talmudic law.

Christian theologians and philosophers continued the Hebrew conceptions of a divine lawgiver. Indeed, even in the early centuries of Christianity it is not hard to find statements which imply laws for non-human Nature. For example, the writer Arnobius (*c.* A.D. 300), arguing that Christianity is nothing monstrous, says that since its introduction there have been no changes in 'the laws originally established'. The elements (of the Greeks) have not changed their properties, the structure of the machine of the universe (the astronomical system) has not dissolved. The rotation of the universe has not altered, the sun has not cooled, the changes of the moon, the turn of the seasons, the succession of long and short days have neither been stopped nor disturbed; it still rains, seeds still germinate, and so on and so on. Yet here we remain at the stage before a very sharp separation between human natural law and non-human laws of Nature had come about. In the early Christian centuries there are many texts which show this quite clearly: in the *Constitution* of Theodosius, Arcadius and Honorius of 375 there is a passage forbidding anyone to practise fortune-telling on pain of punishment for high treason: 'it is impious to tamper with the principles which keep the secret laws of Nature from men's eyes'. A second text, this time of the Roman jurist Ulpian, appears in the sixth-century Justinian *Corpus Juris Civilis* (Corpus of Civil Law), where we find:

> 'Natural law is that which all animals have been taught by Nature; this law is not peculiar to the human species, it is common to all animals which are produced on land or sea, and to fowls of the air as well. From it comes the union of man and woman called by us matrimony, and therewith the procreation and rearing of children; we find in fact that animals in general, the very wild beasts, are marked by acquaintance with this law.'

Historians of jurisprudence are at pains to explain that this never had any influence on subsequent legal thinking, but even if that be so, it was accepted by mediaeval writers and commentators, and very clearly expresses the idea of animals as semi-juristic individuals obeying a code of laws laid down by God. At this point, then, we come very close to the idea of laws of Nature as the divine legislation which all matter, including animal life, obeys.

As the Christian centuries moved on, it was inevitable that natural law should come to be identified with Christian morality. St Paul had clearly recognised this, and in the early fifth century St Chrysostom had seen in the Ten Commandments a codification of natural law. With the famous *Decretum* on ecclesiastical law compiled by the Benedictine monk Gratian in 1148, the identification was complete. It was, moreover, a universal mediaeval belief that commands of princes contrary to natural law were not binding on their subjects and could be resisted. This was a doctrine that bore much fruit at the time of the rise of Protestantism and the beginnings of modern European democracy, and one which corresponds very closely with the Confucian doctrine of Mêng Kho (fourth century B.C.) that subjects have a right to dethrone a ruler who ceases to act according to *li*, a fact not lost on European social thinkers who read Latin translations of the Chinese classics made after 1600 by the Jesuits.

The systematisation of all this came, of course, in the thirteenth-century work of Thomas Aquinas who pictured four systems of law: the *lex aeterna* governing all things at all times; the *lex naturalis* governing all men; and the *lex positiva* laid down by human lawgivers, which he divided into two sections, the *divina* of ecclesiastical law inspired by the Holy Spirit working through the Church, and the *humana* of law enacted by princes and legislatures. Yet he also said:

> 'Every law framed by man bears the nature of a law only in the extent to which it is derived from the Law of Nature. But if on any point it is in conflict with the Law of Nature, it at once ceases to be a law; it is a mere corruption of law.'

This is a very close parallel to what the Confucians urged (in other terms) against the Legalists. If *fa* were contrary to *li*, it must be false *fa*.

When this synthesis of Aquinas was dynamited by the Reformation, natural law began to undergo its greatest development, and a basis of universal human reason was substituted for the former basis of divine will. A secularisation of natural law, a transposition of it from the ecclesiastical to the civil realm, accompanied the rise of nationalism from 1500 onwards. This lived on in many forms, and just as it was thought to have had its origin among the merchants at Rome, so in the seventeenth century it returned to a commercial milieu: foreign merchants, it was said, came within the king's jurisdiction, so engendering an attitude that was to lead in due course to the principle of international law.

The acceptance of law in Renaissance natural science

But what of the scientists and their laws of Nature? In the seventeenth century, the time of the brilliant chemist Robert Boyle and of the genius of Isaac Newton, the concept of laws of Nature 'obeyed' by chemical substances and planets alike appears in a fully developed form. How did this happen? How had this concept reached full maturity? And at what points did it differ from the synthesis of Aquinas and the 'scholastic' philosophers who succeeded him? Certainly it seems that the first use of the expression 'law of Nature' in its scientific sense occurs in 1665 in the first volume of the early scientific journal called the *Philosophical Transactions of the Royal Society of London*, and within thirty years it had become a commonplace, appearing naturally even in works of art – for instance, Dryden's translation of Virgil's *Georgics* – as well as in those on science.

Investigation shows that the idea was present among the philosophers and scientific thinkers of the early seventeenth century, a time when there seems to have been a parallel development of secularised natural law based on human reason, and of the mathematical expression of the empirical laws of Nature. Yet the first thinker to draw out the laws of non-human nature from Aquinas' *lex aeterna* seems likely to have been

Giordano Bruno (1548–1600). Conceiving of the universe as an organism where inorganic as well as organic bodies were animate, he said that God is to be found in the 'inviolable and unbreakable law of Nature', and this even though his thought is, in general, rather Chinese in style. Certainly there is no doubt that even if Bruno was not the first fully to appreciate the concept scientifically, the turning point in Western thought occurred between Copernicus (1473–1543) and Kepler (1571–1630). Copernicus still spoke of symmetries, harmonies and motions when discussing the movements of the planets; he never mentioned 'laws'. Nor, interestingly enough, did Galileo (1564–1642) ever use the expression 'Laws of Nature', even though it was he who published in 1638 *Discorsi e Dimostrazioni Matematiche intorno à due nuovo scienze* (Discourses and Mathematical Demonstrations of Two New Sciences), a book which was the beginning of modern mechanics and mathematical physics. This was true, too, of his contemporaries, except for Kepler. Yet although Kepler discovered the three empirical laws of planetary motion, it was not, strangely enough, in connection with these that he used the term 'law'; he spoke of it when discussing quite another matter, the principle of the lever, and then used it as if it meant measure or proportion. Yet in 1546 Georgius Agricola, the metallurgist and mining engineer, wrote in his *De Ortu et Causis Subterraneorum*, the first Western book on physical geology:

> 'But what proportion of "Earth" is in each liquid from which a metal is made, no mortal can ever ascertain, or still less explain, but the one God has known it, who has given sure and fixed laws to Nature for mixing and blending things together.'

Here was a definite statement at last, and from chemistry, not astronomy.

The concept spread rapidly, and by the time we come to the philosopher and scientist René Descartes (1596–1650) we find the idea of the laws of Nature as fully developed as it was later to be in the works of Boyle and Newton. In his *Discours de la Méthode* of 1637, Descartes speaks of the 'law which God has put into Nature', and in the *Principia Philosophiae* seven years later, he concludes by saying that what had been discussed in the book was 'what must follow from the mutual impact of bodies according to mechanical laws, confirmed by certain and everyday experiments'. So also the philosopher Benedict de Spinoza (1632–77), in his *Tractatus Theologico-Politicus*, distinguishes between those laws 'depending on the necessity of Nature' and those 'resulting from human decrees'.

The most fundamental problem is, however, why, after so many centuries of existence as a commonplace in European theology, the idea of 'laws of Nature' attained a position of such importance in the sixteenth and seventeenth centuries. How was it that, in the modern period, the idea of God's reign over the world shifted from the exceptions in Nature – comets, monsters and the like – to become the rules? The answer seems to be that since the idea of a reign over the world originated by a transference of earthly rulership into the divine realm, we must look to accompanying social developments in order to reach an understanding of the change. If we do this, we at once see that the decline and disappearance of feudalism and the rise of the capitalist state caused a distintegration of the power of the nobility and a great increase in the power of royal authority. The way this happened in Tudor England and in eighteenth-century France is well known, and just when Descartes was writing, the English Commonwealth was taking the process even further, towards an authority which was centralised but no longer royal. If it is reasonable, then, to relate the rise of the Stoic doctrine of Universal Law to the period of the rise of the great monarchies following the death of Alexander the Great, it seems equally reasonable to relate the rise of the concept of laws of Nature at the Renaissance to the appearance of absolute royal power at the end of feudalism and the beginning of capitalism. It appears to be no mere chance that Descartes' idea of God as the legislator of the universe was developed no more than forty years after the political philosopher Jean Bodin had published his theory of sovereignty. Yet this brings us face to face with the paradox that in China, where imperial 'absolutism' covered an even longer period, we hardly meet with the idea at all. This we shall now examine.

Chinese thought and the laws of Nature

In the West between the time of Galen, Ulpian and the Theodosian Constitution on the one hand, and that of Kepler and Boyle on the other, the twin ideas of a natural law, common to all men, and of a body of laws of Nature, common to all non-human phenomena, had become completely differentiated. In China, however, the development of thought on natural law and the laws of Nature differed markedly from what happened in Europe.

In China, as we saw earlier, the theories of Yin and Yang and the Five Elements endured for a very long time. The Taoist thinkers, profound and inspired though they were, never developed any idea of laws of Nature. This may have been because of their immense distrust of the powers of reason and logic. They had, it is true, an appreciation of the relative nature of all things, and the subtlety and immensity of the universe, but while they were groping for what we may call a world-picture of the kind Einstein was later to draw in the West, they did so without laying the right foundations for a Newtonian one. By that path science could not develop. It was not that the Tao, the cosmic order of things, did not work according to system and rule, but that the tendency of the Taoists was to regard it as inscrutable. It would not, perhaps, be going too far to say that this was the one reason why, when the care of Chinese science over the centuries was consigned to them, this science remained on a mainly empirical level. Moreover, it is not irrelevant that their social ideals had less use for positive law than those of any other school. Seeking to go back to primitive tribal collectivism, they had no interest at all in the abstract laws of any lawgivers.

The late Mohists and the Logicians, on the other hand, strove mightily to perfect logical processes, and took the first steps in applying them to zoological classification and to the elements of mechanics and optics, but their movement failed. Why this happened we do not know for certain, but perhaps it was because the Mohists' interest in Nature was too strongly bound up with their practical aims in military technology; at all events their school had few survivors after the upheavals of the first unification of the empire. They did use the term *fa*, but it meant 'causes', rather of the kind discussed by Aristotle ('formal', 'efficient', 'material' and 'final'), and they seem to have approached no nearer to laws of Nature than did the Taoists.

When we come to the Legalists and the Confucians we enter the realm of pure social interest, for neither had any curiosity about Nature outside man himself. The Legalists laid all their emphasis on positive law, on *fa*, which was to be the pure will of the lawgiver irrespective of what morality might say, and quite capable of running contrary to it if the well-being of the state should so require. Yet it was a law that was precisely and abstractly formulated. Against this the Confucians adhered to the body of ancient custom, usage and ceremonial, which included all those practices, such as filial piety, which countless generations of Chinese had instinctively felt to be right. This was *li*, and we may equate it with 'natural' law. In other words, *li* was the sum of all the folk outlook and thinking with its ethical sanctions lit by consciousness. What is more, it seemed necessary that this right behaviour should be taught by paternalist magistrates rather than enforced. Moral exhortation was better than legal compulsion. Confucius had said that if the people were given laws and levelled by punishments, they would try to avoid the punishments without any sense of shame in doing so, whereas if they were 'led by virtue' they would spontaneously avoid disputes and crimes. Good customs are more flexible than formulated laws, and can prevent disturbances before they arise, whereas law can operate only after they have arisen. As the *Li Chi* (Record of Rites) puts it, using symbolism taken from hydraulic engineering, good customs are like dykes and embankments that prevent the onset of flood conditions. Here one can see the point of view which came to dominate Chinese thinking once the Confucians had overthrown the Legalists: correct behaviour in accordance with *li* always depended on the circumstances, so to publish laws beforehand was an absurdity since they could never take into account the complexity of actual events. This is why the codified law was restricted to purely criminal provisions.

While it is convenient to draw the contrast between *fa* and *li*, the earliest form of distinction was between *fa* and *i*, a term usually translated 'justice', but which originally meant 'that which seemed just to the natural man'. 'Laws [should] arise out of justice [*i*], and justice arises out of the common people, and must correspond with what they have at heart', says the *Wên Tzu* (Book of Master Wên) from Han times, and this is the Confucian view: law cannot exist without ethical sanction. The Legalists held just the opposite.

The distinction between *i* and *fa* was remembered throughout Chinese history. One might say, indeed, that *i* stood behind *li* as its justification, its inward and spiritual grace. In the Thang (seventh to tenth centuries A.D.) cases were in fact judged first on the code (*lü*), next on the *li* by reference to Confucian classical texts dealing with behaviour which was right ethically or right by custom, and then finally on *i*. An example occurs in the writings of the poet Pai Chü-I. A man's wife was married to him for three years without bearing a child, and his parents wanted to have her divorced. According to the *Li Chi* of early Han times they were justified, but the wife pleaded that she had no other home to go to. Judgment was that although *li* permitted such a divorce, *i* made it impossible on overriding grounds of humanity. Thus a definite clash could occur between what may be called a lower and a higher conception of natural law. However, in Thang times the main clash was between *lü* and *li*, particularly in cases of vendettas, such actions being forbidden by *lü* and enjoined by *li*. In Sung times things changed and the difficulties that arose were between *lü* and imperial edicts, since these often authorised heavier penalties than the code permitted. But the important thing for us is that *i* was even more closely tied to human-heartedness than was *li*, and that neither could readily be extended to the non-human world.

One can see that *fa* could be applied to laws of Nature, but the significant fact is that this was not done until quite recent times – or only on very rare occasions. Indeed, the only absolutely certain case we know in all ancient and mediaeval Chinese literature occurs in the *Chuang Tzu* (Book of Master Chuang) of the third century B.C., where Chuang Chou praises the silence of the universe:

> Heaven and Earth have the greatest beauty, but they are silent,
> The four seasons have manifest *fa*, but they do not discuss them,
> The ten thousand things have perfect intrinsic principles of order, but they do not talk about them.

But does *fa* definitely mean 'laws' here? The answer is probably not, for from a very long way back, *fa* also meant 'model' and 'method' and these might prove better translations in this case.

The phrase 'thien fa' and the word 'ming'. There is no doubt, however, about the meaning of *fa* when it occurs in the phrase *Thien fa*, the 'Laws of Heaven'. Here it certainly means juristic natural law, and is something like *Li*. And in fact as early as 515 B.C. a feudal leader is reported as saying, 'If you, my kinsmen by birth and marriage, will rally round me according to the Law of Heaven (*Thien fa*).' Yet this is not a reference to a law of Nature in the scientific sense: it concerns human affairs and human society. The Greeks had a close parallel, for in Plato where the words say 'law of Nature' they refer also to a question of human affairs, in this case the 'natural right of the stronger'.

In this realm of ideas it is possible to find many expressions in which Heaven is said to give commands, of which *Thien ming* is almost a commonplace example, particularly in the hands of certain writers like Tung Chung-Shu, who inclined to greater personalisation of Heaven than the majority of scholars. *Ming* (命), decree, is nothing but the development of an ancient graph of a mouth, a tent and a person kneeling (𠇷) and in use refers to heavenly commands to man: thus Tung Chung-Shu writes, 'Heaven, when it constituted man's nature, commanded him to practise love and righteousness' (*Thien chih wei jen hsing ming, shih hsing jen i*). Again a human application, but what we are trying to gain is a glimpse of Heaven commanding non-human things to behave as they do; commanding the stars to rotate nightly in the sky, for instance, but this, apparently, *Thien ming* never does. And neither did *thien fa* nor *li* apply outside human society.

There were a few cases where *li* was extended to cover the behaviour of absolutely everything in the universe, but the use then was essentially poetical. Perhaps the most typical example of this, where *li* is used as 'heavenly', occurs in the *Li Chi* of the first or second centuries B.C.

> From all this it follows that *li* has its origin in the Great
>   Unity.
> This, differentiating, became Heaven and Earth.
> Revolving, it became the Yin and the Yang.
> Changing, it manifests itself in the Four Seasons,
> Dispersing, it appears in the form of the gods and spirits.
> Its revelations are called Destiny.
> Its authority is in Heaven.

And the writer adds that, while *li* is rooted in Heaven, its movement reaches to the Earth. All this amounted to, then, was to say that, in some way or other, human moral order had superhuman (but not necessarily supernatural) authority. Such a conviction did not raise the question of the control of non-human Nature.

Sometimes *fa* alone seems to be applied to mathematical or natural regularities, but a closer look shows that it refers only to fixing measuring standards by decrees of positive law. Thus the *Yin Wên Tzu* (Book of Master Yin Wên), probably of Han times, says:

> 'Of the law there are four kinds. The first is called the
> immutable law [for example, that which governs the relations
> of] prince and minister, superior and inferior. The second is
> called the law which adjusts the customs of the people [for
> example, that which governs the relations of] the capable and
> the rustic, likeness and unlikeness. The third is called the law
> which governs the masses [for example, that which bestows]
> honours and rewards, punishments and fines. The fourth is
> called the law of correct balance [for example, that which has
> to do with] calendrical science, acoustics, the degrees of the
> circle, balances and weights.'

Here the first law is certainly juristic natural law, and the second is similar, connected with the natural processes whereby people find their level in society according to their abilities. The third covers both natural and positive law. The borderline with true laws of Nature could be said to be approached with the fourth law, but only if we consider, for instance, that planetary motions must be kept regular if they are to be used for calendar determinations, or pitch-pipes of a pre-determined size if they are to emit definite notes, and so on. But that does not seem to be meant here. What the writer most probably had in mind was the action of the ruler in promulgating laws about those weights and measures, festivals and other dates, etc., that his experts recommended to him.

The passage quoted goes on to say that the sages modelled themselves on Heaven and Earth, and that kings should do so as well. The obvious conclusion, then, is that we have a poetical or metaphorical derivation of human law, the qualities of which were thought of as mirroring certain desirable qualities seen in non-human Nature. But the paradox remains that it should never have occurred to anyone as odd that law could be derived from where no law existed. Clearly, an intuitive idea of the emergence of novelty at the human level was extremely strong in Chinese thought.

The words ' Li ' and ' tsê '. So far, then, we have not found in Chinese thought any clear evidence of the idea of law in the strict sense of the natural sciences. But what of the more advanced ideas of the Neo-Confucians of the Sung dynasty?

We have already seen (Chapter 14) that Chu Hsi and other thinkers in his group made a great effort to bring all Nature and Man into one philosophical system. The principal concepts they worked with were *Li* and *chhi*. The second corresponded to matter and energy, while the first, though used in a technical sense, was not far removed from the Taoist conception of the Tao as the Order of Nature. But the word *Li* in its most ancient meaning signified the pattern in things – the markings in jade or the fibres in muscle – and as a verb, to cut things according to their natural grain or natural divisions. Thus *Li* acquired the common dictionary meaning of 'principle'. Chu Hsi said:

> ' *Li* is like a piece of thread with its strands, or like this
> bamboo basket... One strip goes this way; another strip goes
> that way. It is also like the grain in the bamboo – on the
> straight it is of one kind, and on the transverse it is of another
> kind. So also the mind possesses numerous principles [*Li*].'

Li, then, is rather the order and pattern in Nature, not a formulated law. However, it is not a dead pattern like a mosaic, but a dynamic pattern as embodied in living things and in human relationships; a dynamic pattern that can only be expressed by the word 'organism'. Yet in an important passage in the *Chu Tzu Chhüan Shu* we find that *Li* is connected with another word *tsê*, a word that is of no small importance. The passage runs:

> ' *Question.* In distinguishing between the four terms Heaven
> [*thien*], Fate [*ming*], the Nature [*hsing*] and *Li*, would it be

correct to speak as follows? In the term Heaven the reference
is to spontaneous naturalness. In the term Fate the reference
is to its flowing through the pervading the universe, and being
present in all things. In the term Nature the reference is to
that complete provision which any specific thing must have
before it can come into being. In the term *Li*, the reference is
to the fact that every event and thing has each its own rule of
existence [*tsê*]. And taking them all together, may it not be
said that Heaven [i.e. the natural universe as a whole] is *Li*,
that Fate is in fact the Nature [i.e. the constitution of a thing
or a man], and that the Nature is in fact also *Li*? Is this not
correct?

'*Answer.* You are right.'

The operative word here is in fact *tsê*, which has been translated as 'rules
of existence', for there is no doubt that Chu Hsi's questioner had in
mind a well-known passage in the early *Shih Ching* (Book of Odes)
(before the sixth century B.C.):

Heaven, in giving birth to the multitudes of the people,
To every faculty and relationship annexed its law (*tsê*)
The people possess this normal nature
And [consequently] love its normal virtue.

This famous verse was quoted by Mencius and referred to again by Chu
Hsi himself, who gives it as his opinion that the 'faculties and
relationships' are the likes and dislikes of human desire; to like that
which is good and to dislike evil is the rule of existence which was
translated here by James Legge as 'law'. In other words, we have to
deal with neutral natural properties on the one hand, and their regular
tendency to behave in a specific manner on the other.

Once again we are certainly in the no-man's-land between scientific
laws and natural law in the legal sense; we are back in those shadowy
regions where the concepts are in a highly undifferentiated state. But
we can make a discovery of much interest if we look at the origin of
tsê. The ancient character (鼎) on bones and bronzes depicts a cauldron
and a knife, or the very act of incising codes on ritual cauldrons (as
described under entry 72 in Table 8 on page 138). The word, however,
underwent various vicissitudes with time, being corrupted into meaning
'cowrie shells', used as the particles 'so', 'then' and 'in that case', and
with secondary uses connected with laws and tariffs, for instance *chhang
tsê*, 'unvarying laws', and *shui tsê*, 'excise tariff'. If this were all,
identification of Chu Hsi's *Li* with the *tsê* of the Odes would certainly

not be convincing, but there is more. There is, for instance, an astro-nomical poem where *tsê* may, it seems, be translated as a rule or law, although one cannot be certain that this is correct. Moreover there are other references which make it seem possible that *tsê* could have been used in this way, but our doubts arise from the fact that there are also definite denials of the application of *tsê* to natural phenomena. In the *Huai Nan Tzu* of the second century B.C. it is said:

> 'The Tao of Heaven operates mysteriously and secretly; it has no fixed shape; it follows no definite rules [*wu tsê*]; it is so great that you can never come to the end of it; it is so deep you can never fathom it'

and again, eight centuries later, Liu Tsung-Yuan wrote:

> 'Heaven has no colour of any kind, no centre and no sides – how can you hope to find its *tsê*?'

Nor are these isolated instances; there are others which are quite emphatic, especially Wang Pi's third-century A.D. commentary on the *I Ching*:

> 'The general meaning of the Tao of "Kuan" is that one should not govern by means of punishments and legal pressure, but by looking forth one should exert one's influence [by example] so as to change all things. Spiritual rule is without form and invisible. We do not see Heaven command the four seasons, and yet they never swerve from their course.'

This is perhaps the most illuminating passage of all, for here we have a flat denial of any conception of orders from a celestial lawgiver. The thought is extremely Chinese: universal harmony comes about by spontaneous co-operation, not celestial fiat; by things following the internal necessities of their own natures. This Chinese doctrine of organism was, indeed, deeply rooted in Neo-Confucianism; as Chang Tsai put it in the eleventh century when referring to the heavens: 'All rotating things have a spontaneous force [*chi*] and thus their motion is not imposed on them from outside.' For our part, we should, therefore, be much mistaken to view *tsê* in anything like the sense of laws of Nature as Newton, for instance, would have conceived of them.

The Chinese denial of a celestrial lawgiver. It is worth following the question of the denial of a celestial lawgiver just a little further, for the affirmation that Heaven does not *command* the processes of Nature to follow their regular courses is linked with the root idea of Chinese thought – *wu wei*, non-action or unforced action. The legislation of a lawgiver would, however, be *wei*, forcing things to obedience. Indeed, it is not difficult to find passages which confirm the conception of Heaven acting according to *wu wei*. It runs through the fourth-century B.C. *Tao Tê Ching* (Canon of the Virtue of the Tao), where we find the significant statement that though the Tao produces, feeds and clothes the myriad things, it does not lord it over them, and asks nothing of them. This idea is, of course, a Taoist commonplace, echoed in the eleventh century by Chhêng Hao when he wrote: 'The laws of Heaven are wordless but they keep faith, divine law has majesty untinged with wrath.' A sublime sentiment, yet one that is obviously incompatible with the concept of a personal celestial lawgiver.

'*Li*' and '*tsê*' in Neo-Confucianism

It would be valuable for the history of Chinese scientific thought to concentrate attention on other occurrences of the phrase *Thien tsê*, yet even as far as studies have gone at present, it is clearly not a common phrase. *Tsê* itself seems to represent a borderline conception. Certainly it always had a legal side to it, and a human one too, but although applied occasionally in a scientific sense, such a use does not seem to have caught on. Chu Hsi would certainly have pondered on the classical texts of Chang Hêng and known of the rare uses of *tsê* in contexts that brought it close to the meaning of scientific law. But how far his idea of *Li* involved the conception of laws of Nature can hardly be assessed until more is known of the emphasis of the passages he is likely to have had in mind. All the same, there is one feature of the crucial dialogue already given on pp. 296/7 which suggests that laws of Nature in the sense of general scientific laws of behaviour were not meant: this is the sentence in italics 'every event and thing has each *its own* rule of existence'. It is not said that every event and thing obeys general laws or rules valid for many other similar events and things, and the thought here is one that is far more applicable to individual events, and things as organisms. There is no absolute contradiction, certainly, for it is a matter of difference in emphasis, but it agrees with Wang Pi's statement about the general meaning of the Tao (on the previous page).

A further insight into what the Neo-Confucians meant by *Li* and *tsê* may be found by looking into the *Pei-Chhi Tzu I* (Philosophical Glossary of Neo-Confucian Technical Terms) written by Chhen Shun, an immediate pupil of Chu Hsi about the time of the latter's death (A.D. 1200). Tao, he says, is broader than *Li*, but *Li* is more profound. *Li* is formless, but yet a pattern and organisation which is *a natural and inescapable law*. The *Li*, he goes on, permeates non-human things; it is that universal *Tao Li* which is common to Heaven and Earth, and all human beings, and all things. *Li* is what organises substance, and *i* is what organises functions. *Li* permeating things is the natural inescapableness of them; *i* is how to handle, direct or administer this *Li*.

In essence, then, *Li* is indeed the 'principle of organisation' as we called it when discussing the Neo-Confucians. There is law implicit in it, but it is a law to which parts of wholes have to conform by virtue of their very existence as parts of wholes. And this is true whether they are human parts or non-human parts. This 'law' arose not by decree of a universal Controller but directly out of the nature of the universe. *Li* is no fortuitous concourse of atoms obeying statistical laws of their own either; it is in no way connected with the patterns of chance. The cosmic order is whole and unchanging; it is a Great Pattern in which lesser patterns are included, and the 'laws' which are involved are *intrinsic* to these patterns, an integral part of them, not extrinsic to them or dominating them as the laws of human society dominate men. The laws of the Neo-Confucian philosophy of organism would thus be internal to individual organisms at all levels, just as in later Western civilisation, it was felt that the laws of an ideal state should be written not on tablets but in the hearts of its citizens.

We must conclude, then, that 'law' was understood in an 'organic' sense by the Neo-Confucian school: law in the sense used in describing the mathematical universe of Newton was either completely absent from their definition of *Li* or, at the most, played a very minor part. The main component was 'pattern', including pattern living and dynamic to the fullest extent; in other words 'organism'. In this philosophy of organism, all things in the universe were included: thus Heaven, Earth and Man all have the same *Li*.

What exactly all this means is important for our argument and is expounded in a long passage by Chu Hsi. Just like Chuang Chou 1400 years before, he asserts that *Li* runs through all things in the universe, that the universe is orderly and, in a sense, rational. But it is not thereby intelligible in the scientific as opposed to the philosophical sense, and not necessarily following rules capable of being formulated in a precise and abstract way by man. Moreover, Chu Hsi expresses what amounts to a conception of ethics in terms of levels of organisation. 'Inorganic' objects have their place, relatively low, in the overall pattern, and ethical and moral phenomena, properly so called, only begin to appear when a sufficiently high level of organisation is reached. They appear first, incompletely and one-sidedly, in animals, and only reach their full expression in man. In fact Chu Hsi claims that moral concepts are not applicable to inorganic objects, yet, like the Taoists before him, he still cannot find terms other than 'natural endowment' and even 'mind' when he wants to describe something like the properties of chemical substances. Certainly one of the things he tried to do was to grope his way towards a classification of chemical properties; in the examples he gives he undoubtedly steps into that long avenue which led to the inorganic and organic chemistry of our own time, and yet the necessary conceptual language was still wanting.

We seem then to be, in the latter part of the twelfth century, in the presence of a point of view similar to that which Ulpian expressed in Europe nearly a millennium earlier. Yet there was one profound difference. Ulpian had spoken uncompromisingly of 'law', but Chu Hsi relies chiefly on a technical term the primary meaning of which is 'pattern'. For Ulpian all things were 'citizens' subject to universal law: for Chu Hsi all things were 'dancers' in a universal pattern. We can say, therefore, that nothing more than traces of the concept of laws of Nature can be found in the greatest of the Chinese philosophical schools, the Neo-Confucians of the Sung.

Order which excludes law

At the end of this investigation we have to conclude that not one of the words in ancient and mediaeval Chinese texts which have tempted translation as the 'laws of Nature' gives us any right to do this. The Chinese outlook ran along quite different lines: the Chinese notion of Order positively excluded the notion of Law. Yet so unconscious has the idea of laws of Nature been among Europeans that many Western scholars of Chinese have unsuspectingly read the word 'law' into texts when in fact there is no Chinese word there to justify it. For instance, one translator of the *Yen Thieh Lun* (Discussion on Salt and Iron) writes, 'The Tao hung its laws in the Heavens and spread its products on the earth', but all the text actually says is 'The Tao [the Order of Nature] is hung up [manifest] in the Heavens', and this is no isolated case. Of course free translations are always more attractive than literal ones, but they are liable to suffer from the unconscious intellectual background of the translator, and there are occasions when this matters a great deal.

Judicial trials of animals: contrasting European and Chinese attitudes. Before concluding it may be as well to glance at a striking illustration of the difference in outlook between China and Europe on law and on Nature. As the reader is probably well aware, during the European Middle Ages there were considerable numbers of trials and criminal prosecutions of animals in courts of law, followed frequently by capital punishment. The trials stretched from the ninth to the nineteenth centuries, reaching a peak in the sixteenth, the time of the well-known witch-mania, and they fall into three categories. First, the trial and execution of domestic animals for attacking human beings (e.g. pigs for devouring infants); secondly the excommunication or ecclesiastical cursing of plagues or pests of birds and insects; thirdly, the condemnation of freaks of nature (e.g. the laying of eggs by cocks). It is the last two that are the most interesting for our present theme, but one example must suffice. At Basle in 1474 a cock was sentenced to be burnt alive for the 'heinous and unnatural crime' of laying an egg. What had actually happened in this and similar cases was probably a sex reversal, where the plumage of the hen closely approximates to that of a cock with the result that it would be presumed to be a cock, at a time before the understanding of endocrinology and the anatomy of the sex organs. At all events, whatever the cause, one of the reasons for the alarm involved was probably that cocks' eggs were an ingredient in witches' ointments, and another, even worse, was the possibility (or probability) that a cock's egg would hatch into the feared cockatrice or basilisk, fabled to be capable of killing with a glance.

The interest in the story for us is that such trials would have been absolutely impossible in China. The Chinese were not so presumptuous as to suppose that they knew the laws laid down for non-human things so well that they could proceed to indict an animal at law for transgressing them. On the contrary, the Chinese reaction would undoubtedly have been to treat these rare and frightening events as 'reprimands from Heaven', and it was the emperor or the provincial governor whose position would have been endangered, not the cock. Indeed, we can quote chapter and verse for such an interpretation. In the long Record of (Derangements of) the Five Elements in the *Chhien Han Shu* (History of the Former Han Dynasty) there are several references to sex reversals in poultry and in Man. These were classified as 'green misfortunes' and thought of as connected with the activities of the element Wood. They foreboded serious harm to the rulers in whose dominions they occurred.

There were, then, fundamental differences between what may be called the dominant attitudes of the Chinese and Western civilisations. Behaviour characteristic of the opposite one did occur, however, in less deep-rooted matters. Thus in late Chinese folklore there are stories of animals being brought before magistrates' courts, but what is significant is that the Chinese emphasis was different, for the stories usually concern such themes as the repentance of tigers for killing men. They are patently Buddhist in inspiration, belonging to the first class of legal prosecution mentioned above, but the important cases for the present argument are, of course, those of the third class, where no harm had been done to man and where, as we have seen, the Chinese attitude was primarily one of resignation to a heavenly visitation. Nevertheless there were, on occasions, some attempts to take active measures. In 716 Yao Chhung, a famous minister of the Thang, urged in a memorial to the emperor that plagues of locusts were perfectly natural and not the result of a 'phenomenalist' reprisal on the part of Heaven. He therefore organized nation-wide counter-measures. Again, a century before, the Emperor Thai Tsung publicly ate a plate of fried locusts to demonstrate that they were not something sacred sent from Heaven as a punishment. But these were practical reactions, not prosecutions at law.

Dominance psychology and excessive abstraction. Pondering over the differing Eastern and Western conceptions of law in relation to the living world, it may occur to the reader that some difference in emphasis might arise according to whether Man has to do chiefly with the animal or the vegetable kingdoms. After all, contrasting attitudes do originate in different environments. In a pastoral setting the shepherd and cowherd beat their beasts; they take an active attitude of command over their flocks and herds, and even God is imagined as a 'Good Shepherd' leading his flock to satisfying pastures. But the shepherd is not far from the legislator, and pastoral dominance over animals consorts well with legislation over things as well as men. Maritime usages also strongly reinforce this command-psychology, for the safety of all on board a ship requires unquestioning obedience of the crew to the experienced captain. Hence laws in Nature could have been – and were – derived from attitudes that accepted the masteries of shepherds and sea-captains as well as kings.

But when man has to do primarily with plants, as in predominantly agricultural civilisations, the psychological conditions were quite different. Often the less he interferes with the growth of his crops the better; until the harvest he does not touch them, leaving them meanwhile to follow their Tao which leads, in due course, to his benefit. So the conception of *wu wei* – no action contrary to Nature – is profoundly in accord with the agricultural peasant life. Indeed in Mencius there is a famous story underlining this:

> 'Let us not be like the man of Sung. There was once a man of Sung, who was grieved that his corn was not longer, so he pulled it up. Returning home, looking very stupid, he said to his people, "I am tired today. I have been helping the corn to grow long." His son ran to look at it, and found the corn all withered.'

Agricultural civilisations would not, therefore, be expected to show dominance psychology, or the notion of a divine legislator which is, perhaps, associated with it. If, indeed, this notion first began in Babylonia, it was no doubt because the ancient economy of the fertile crescent was a mixed one, and much of its spread was due to that pre-eminently pastoral people, the Hebrews.

Another point of difference between Chinese and European conceptions of law involves not biology but mathematics. In contrast with the Greek genius for geometry, the Chinese had a gift for arithmetic and algebra. Now there is something suspiciously similar between the abstractness of geometry and the abstractness of Roman law, where an agreement between two persons was deemed to have no possible bearing on a third. But for Chinese law such an abstractness is inconceivable: no agreement could be considered in isolation from the practical circumstances surrounding it, from the positions and obligations of persons in society and the effects it might have on other persons. Just as Greek geometry dealt with pure and abstract figures, the size of which was quite immaterial once the principle had been accepted, so Roman law dealt with codified abstractions. But the Chinese preferred to think only on concrete numbers (though in algebra they might not be any particular numbers), and, similarly, of concrete social circumstances when it came to matters of law.

Comparative philosophy of law in China and Europe. From what has been said earlier, it is clear that in China the theories of the Yin and Yang or the Five Elements had the same status as the early scientific theories of the Greeks. What went wrong with Chinese science was its ultimate failure to develop, out of these theories, forms more adequate to the growth of practical knowledge and, in particular, its failure to apply mathematics to the regularities to be found in Nature. But for that situation, the specific nature of the social and economic system must be held largely responsible; differences in apprehension of Nature as such cannot, it seems, explain the differences between Chinese and European conceptions of law.

In Europe natural law may be said to have helped the growth of natural science because natural law had such great universality. But in China 'natural law' was never thought of as law and took a very social name, *Li*. It was therefore hard to think of any 'law' as applicable outside human society, even though, relatively speaking, *li* was much more important for society than natural law was in Europe. When order or pattern were visualised as running through the whole of Nature, it was generally not as *li* but as the Tao of the Taoists or the *Li* of the Neo-Confucians; conceptions, in fact, which had no juristic content.

Again, in Europe, positive law may be said to have helped the growth of natural science because of its precise formulation. This encouragement came about because of the idea that to the earthly lawgiver there corresponded a celestial one in Heaven, whose writ ran wherever there were material things. In order to believe that Nature was rational and intelligible the Western mind found it congenial to suppose the existence of a Supreme Being, himself rational, who had put it there. The Chinese mind did not think in these terms at all. Imperial majesty corresponded not to a legislative creator but to a Pole Star, to the focal point of a universal ever-moving pattern and harmony not made with hands, even those of God. And the pattern was rationally intelligible only because it was embodied in Man himself.

Chinese conceptions of deity. Here is not the place to investigate the Chinese conceptions of God, although there are a couple of points which must be mentioned because they are relevant to what we have been discussing. The first is that it is clear that the de-personalisation of God in ancient Chinese thought took place so early, and went so far, that the conception of a divine celestial lawgiver imposing ordinances on non-human Nature never developed. The second is that the highest spiritual being ever known and worshipped in China was not a creator in the sense meant by the Hebrews and the Greeks.

It was not that there was no order in Nature for the Chinese, but rather that it was not an order ordained by a rational personal being. Hence there was no conviction that rational personal beings would be able to spell out, in their lesser earthly languages, the divine code of laws which He had previously decreed. The Taoists indeed would have scorned such an idea as too naïve for the subtlety and complexity of the universe as they saw it. As we have seen, the Chinese had another faith: the universal order was intelligible to human beings because they themselves had been produced by it. They were its highest component pattern.

What is extremely interesting is that modern science, which has found it possible, and even desirable, to dispense completely with the hypothesis of God as the basis for the laws of Nature, has returned in a sense to the Taoist outlook. This is what accounts for the strangely modern ring in so much of the writing of that great school. But historically the question remains whether natural science could ever have reached its present stage of development without passing through a 'theological' phase.

Conclusions

To sum up, then, we may say that the conception of the laws of Nature did not develop from Chinese juristic theory or practice. The reasons were, first, that the Chinese acquired a great distaste for precisely formulated abstract codified law, a distaste that arose from their unhappy experiences with the Legalists during the transition period from feudalism to bureaucracy. Secondly, once the bureaucratic system had firmly settled in, the old conceptions of *li* proved more suitable than any others for Chinese society, and so the element of natural law became relatively more important in China than in Europe. But the fact that so little of this natural law was expressed in formal legal terms, and also that it was overwhelmingly social and ethical, made any extension of its sphere of influence to non-human Nature impossible. Thirdly, the idea of a supreme being, though certainly present from the earliest times, soon lost all personal and creative qualities. As a result, the development of the idea of precisely formulated abstract rational laws which could be deciphered and re-stated simply because there had been a rational Author of Nature did not occur.

The Chinese world-view depended on a totally different line of thought. The harmonious co-operation of all beings arose because they were all parts in a hierarchy that formed a cosmic pattern. They obeyed the dictates of their own natures, not the orders of some superior authority. Modern science and the philosophy of organism have come back to this wisdom, fortified by a new understanding of cosmic and biological evolution.

Lastly, there was always the environment of Chinese social and economic life which moved straight from feudalism to bureaucratism. This was bound to affect Chinese science and philosophy at every step. Had these conditions been more favourable to science the opposing intellectual factors considered in this book might well have been overcome. All we can say of that science of Nature which would then have developed is that it would have been profoundly organic and non-mechanical. But what manner of disciplines the sciences of ancient and mediaeval China actually were is a subject for the next and subsequent volumes of this abridgement.

BIBLIOGRAPHY

Andersson, J. G., *Children of the Yellow Earth*, Kegan Paul, London, 1934.

Anon., *An Outline of the History of China*, Foreign Languages Press, Peking, 1958.

Bagchi, P. C., *India and China; a thousand years of Sino-Indian Cultural Relations*, Hind Kital, Bombay, 1944, 2nd edn 1950.

Balazs, E., *Chinese Civilization and Bureaucracy; Variations on a Theme*, Yale University Press, New Haven, 1964; reprint in reduced format, 1968.

Beasley, W. G. and Pulleyblank, E. G. (ed.), *Historians of China and Japan*, Oxford University Press, London, 1961.

Bodde, D., *China's First Unifier, a study of the Ch'in Dynasty as seen in the life of Li Ssu (280 to 208 B.C.)*, Brill, Leiden, 1938.

Cameron, N., *Barbarians and Mandarins; Thirteen Centuries of Western Travellers in China*, Walker and Weatherhill, New York and Tokyo, 1970.

Carter, T. F., *The Invention of Printing in China and its Spread Westward*, Columbia University Press, New York, 1925; revised editions 1931 and 1955.

Chang Kuang-Chih, *The Archaeology of Ancient China*, Yale University Press, New Haven and London, 1963.

Chêng Tê-Khun, *Archaeology in China:* vol. 1, *Prehistoric China*, Heffer, Cambridge, 1959; vol. 2, *Shang China*, Heffer, Cambridge, 1960; vol. 3, *Chou China*, Heffer, Cambridge, and University Press, Toronto, 1963; vol. 4, *Han China*, in the press.

Chêng Tê-Khun, *New Light on Prehistoric China*, Heffer, Cambridge, and University Press, Toronto, 1966.

Chhen Hêng-Chê (Sophia H. Chen Zen) (ed.), *Symposium on Chinese Culture*, China Institute of Pacific Relations, Shanghai, 1931.

Chhen Shou-Yi, *Chinese Literature; a Historical Introduction*, Ronald, New York, 1961.

Creel, H. G., *The Birth of China*, tr. into French by M. C. Salles, Payot, Paris, 1937.

Creel, H. G., *Chinese Thought from Confucius to Mao Tsê-Tung*, University of Chicago Press, Chicago, 1953.

Cressey, G. B., *China's Geographic Foundations; A Survey of the Land and its People*, McGraw-Hill, New York, 1934.

Eberhard, W., *Conquerors and Rulers; Social Forces in Mediaeval China*, Brill, Leiden, 1952, 2nd ed. revised 1965.

Eberhard, W., *A History of China from the Earliest Times to the Present Day*, Routledge and Kegan Paul, London, 1950.

Eichhorn, W., *Chinese Civilisation; an Introduction*, Faber and Faber, London, 1969.

Fêng Yu-Lan, *A Short History of Chinese Philosophy*, ed. D. Bodde, Macmillan, New York, 1950.

Fêng Yuan-Chün, *A Short History of Classical Chinese Literature*, tr. Yang Hsien-Yi and Gladys Yang, Foreign Languages Press, Peking, 1958; reprinted 1959.

Fitzgerald, C. P., *China; a Short Cultural History*, Cresset Press, London, 1935.

Geil, W. E., *The Great Wall of China*, Murray, London, 1909.

Giles, H. A., *A Chinese Biographical Dictionary*, 2 vols. Kelly and Walsh, Shanghai and Quaritch, London, 1898; Supplementary Index by J. V. Gillis and Yü Ping-Yüeh, Peiping, 1936.

Goodrich, L. Carrington, *Short History of the Chinese People*, Harper, New York, 1943.

Graham, A. C., '"Being" in Western Philosophy compared with *shih/fei* and *yu/wu* in Chinese Philosophy', with an appendix on 'The Supposed Vagueness of Chinese', *Asia Major*, 1959, **7**, 79.

Graham, A. C., 'The Concepts of Necessity and the *a priori* in Later Mohist Disputation', *Asia Major*, 1975, **19**, 163.

Graham, A. C., 'The "hard and white" Disputations of the Chinese Sophists', *Bulletin of the London School of Oriental and African Studies*, 1967, **30** (no. 2), 358.

Graham, A. C., 'Later Mohist Treatises on Ethics and Logic reconstructed from the *Ta-Chhü* Chapter of the *Mo Tzu* Book', *Asia Minor*, 1972, **17** (N.S.), 137.

Graham, A. C., 'The Logic of the Mohist *Hsiao-Chhü*', *T'oung Pao*, 1964, **51** (no. 1), 1–54.

de Grazia, S. (ed.), *Masters of Chinese Political Thought, from the Beginnings to the Han Dynasty*, Viking, New York, 1973.

Herrmann, A., *Historical and Commercial Atlas of China*, Harvard-Yenching Institute, Cambridge, Mass., 1935; 2nd ed., *An Historical Atlas of China*, ed. N. Ginsburg, with preface by P. Wheatley, Edinburgh University Press, Edinburgh, and Aldine, Chicago, 1966.

Hirth, F. and Rockhill, W. W. (trs.), *Chau Ju-Kua; His work on the Chinese and Arab Trade in the 12th and 13th centuries, entitled 'Chu-Fan-Chi'*, Imperial Academy of Science, St Petersburg, 1911.

Hirth, F., *China and the Roman Orient*, Kelly and Walsh, Shanghai; G. Hirth, Leipzig and Munich, 1885 (photographically reproduced in China with no imprint, 1939).

Ho Ping-Ti, *The Cradle of the East; an Enquiry into the Indigenous Origins of Techniques and Ideas of Neolithic and Early Historic China, 5000 B.C. to*

1000 B.C., Chinese University, Hong Kong, and University of Chicago Press, Chicago, 1975.

Hookham, H., *A Short History of China*, Longmans Green, London, 1969.

Hsü Shih-Lien, *The Political Philosophy of Confucianism*, Routledge, London, 1932.

Hu Shih, *The Chinese Renaissance*, University of Chicago Press, Chicago, 1934.

Hu Shih, *The Development of the Logical Method in Ancient China*, Oriental Book Co., Shanghai, 1922.

Hucker, C. O., *China; a Critical Bibliography*, University of Arizona Press, Tucson, 1962; reprinted 1964, 1966.

Hudson, G. F., *Europe and China; A Survey of their Relations from the Earliest Times to 1800*, Arnold, London, 1931.

Hughes, E. R., *Chinese Philosophy in Classical Times*, Dent, London, 1942 (Everyman's Library, no. 973).

Jacobs, N., *The Origin of Modern Capitalism and Eastern Asia*, University Press, Hong Kong, 1958.

Kaltenmark, O., *Chinese Literature*, tr. from the French (1948) by A. M. Geoghegan, Walker, New York, 1964.

Karlgren, B., *Sound and Symbol in Chinese*, Oxford, 1923; reprinted 1946.

Kroeber, A. L., *Configurations of Culture Growth*, University of California Press, Berkeley and Los Angeles, 1944.

Lattimore, O., *Inner Asian Frontiers of China*, Oxford University Press, London and New York, 1940.

Lattimore, O. and Lattimore, E. (ed.), *Silks, Spices and Empire; Asia seen through the Eyes of its Discoverers*, Delacorte, New York, 1968.

Laufer, B., *Sino-Iranica; Chinese Contributions to the History of Civilisation in Ancient Iran*, Field Museum of Natural History (Chicago) *Publications –* Anthropological Series, 1919, 15, no. 3 (Pub. no. 201) (rev. and crit. Chang Hung-Chao, Memoirs of the Chinese Geological Survey, 1925 (ser. B), no. 5).

Levenson, J. R. and Schurmann, F., *China; an Interpretative History, from the Beginnings to the Fall of Han*, University of California Press, Berkeley and Los Angeles, 1969; reprinted 1971.

Liang, Chhi-Chhao, *History of Chinese Political Thought during the early Tsin [Chhin] Period*, Kegan Paul, London, 1930.

Loewe, M., *Everyday Life in Early Imperial China; during the Han Period, 202 B.C. to A.D. 220*, Batsford, London and Putman, New York, 1968.

Loewe, M., *Imperial China; the Historical Background to the Modern Age*, Allen and Unwin, London, 1966.

Lu Gwei-Djen and Needham, Joseph, 'China and the Origin of (Qualifying) Examinations in Medicine', *Proceedings of the Royal Society of Medicine*, 1963, **56**, 63.

McNair, H. F. (ed.), *China* (collective essays), University of California Press, Berkeley and Los Angeles, 1946.

Mirsky, Jeannette, *The Great Chinese Travellers*, Allen and Unwin, London, 1965.

Moule, A. C., *Christians in China before the year 1550*, Society for the Promotion of Christian Knowledge, London, 1930.

Moule, A. C. and Pelliot, P. (tr. and annot.), *Marco Polo* (A.D. *1254 to* A.D. *1325*); *The Description of the World*, 2 vols., Routledge, London, 1938; reprinted AMS Press, New York, 1976.

Needham, Joseph, *Clerks and Craftsmen in China and the West* (Collected Lectures and Addresses), Cambridge University Press, Cambridge, 1970.

Needham, Joseph, *The Development of Iron and Steel Technology in China*, Newcomen Society, London, 1958.

Needham, Joseph, *The Grand Titration; Science and Society in China and the West* (Collected Addresses), Allen and Unwin, London, 1969.

Needham, Joseph, *Time and Eastern Man* (Henry Myers Lecture, Royal Anthropological Institute 1964), Royal Anthropological Institute, London, 1965.

Needham, Joseph, 'The Translation of Old Chinese Scientific and Technical Texts', art. in *Aspects of Translation*, ed. A. H. Smith, Secker and Warburg, London, 1958, p. 65; Studies in Communication, no. 2; and *Babel*, 1958, **4** (no. 1), 8.

Needham, Joseph and Lu Gwei-Djen, 'Hygiene and Preventive Medicine in Ancient China', *Journal of the History of Medicine and Allied Sciences*, 1962, **17,** 429.

Needham, Joseph and Lu Gwei-Djen, 'The Optick Artists of Chiangsu', *Proceedings of the Royal Microscopical Society* (Oxford Symposium Volume), 1967, **2**, 113.

Needham, Joseph, Wang Ling and Price, D. J. de S., *Heavenly Clockwork; the Great Astronomical Clocks of Mediaeval China*, Cambridge University Press, Cambridge, 1960 (Antiquarian Horological Society Monographs, no. 1).

Nivison, D. S. and Wright, A. F. (ed.), *Confucianism in Action*, Stanford University Press, Stanford, California, 1959.

Radhakrishnan, S., *India and China*, Hind Kital, Bombay, 1947.

Reichwein, A., *China and Europe; Intellectual and Artistic Contacts in the Eighteenth Century*, Kegan Paul, London, 1925, tr. from the German edn, Berlin, 1923.

Sarton, George, *Introduction to the History of Science*, vol. 1, 1927; vol. 2, 1931 (2 parts); vol. 3, 1947 (2 parts); Williams and Wilkins, Baltimore (Carnegie Institution Pub. no. 376).

Schafer, E. H., *The Golden Peaches of Samarkand; a Study of Thang Exotics*, University of California Press, Berkeley and Los Angeles, 1963; Rev. J. Chmielewski, *Orientalische Literatur-Zeitung*, 1966, **61**, 497.

Schafer, E. H., *The Vermilion Bird; Thang Images of the South*, University of California Press, Berkeley and Los Angeles, 1967.

Schafer, E. H., *Ancient China*, Time-Life, New York, 1967.

Sickman, L. and Soper, A., *The Art and Architecture of China*, Penguin (Pelican), London, 1956.

Sigerist, H. E., *A History of Medicine*, 2 vols., Oxford University Press, New York and Oxford, 1951.

Singer, C., *A Short History of Anatomy and Physiology from the Greeks to Harvey*, Dover, New York, 1957.

Singer, C., *A Short History of Biology*, Oxford University Press, Oxford, 1931.

Singer, C., *A Short History of Scientific Ideas to 1900*, Oxford University Press, Oxford, 1959.

Singer, C., Holmyard, E. J., Hall, A. R. and Williams, T. I. (eds.), *A History of Technology*, 5 vols., Oxford University Press, Oxford, 1954–8.

Singer, C. and Underwood, E. A., *A Short History of Medicine*, Oxford University Press, Oxford, 1962.

Smith, B. and Ong Wan-Ko (Wêng Wango), *China; a History in Art*, Harper and Row, New York and London, n.d. (1971).

Tarn, W. W., *The Greeks in Bactria and India*, Cambridge University Press, Cambridge, 1951.

Teggart, F. J., *Rome and China; a Study of Correlations in Historical Events*, University of California Press, Berkeley and Los Angeles, California, 1939.

Têng Ssu-Yü and Biggerstaff, K., *An Annotated Bibliography of Selected Chinese Reference Works*, Harvard-Yenching Institute, Peiping, 1936.

Treistman, J. M., *The Prehistory of China; an Archaeological Exploration*, David and Charles, Newton Abbot, 1972.

Waley, A. (tr.), *The Way and its Power; a study of the 'Tao Tê Ching' and its Place in Chinese Thought*, Allen and Unwin, London, 1934.

Waley, A., *Three Ways of Thought in Ancient China*, Allen and Unwin, London, 1939.

Werner, E. T. C., *A Dictionary of Chinese Mythology*, Kelly and Walsh, Shanghai, 1912.

von Wiethof, B., *An Introduction to Chinese History, from Ancient Times to 1912*, Thames and Hudson, London, 1975.

Willetts, W. Y., *Foundations of Chinese Art; from Neolithic Pottery to Modern Architecture*, Thames and Hudson, London, 1965, revised, abridged and rewritten version, with many illustrations in colour.

Wright, A. F., *Buddhism in Chinese History*, Stanford University Press, Stanford, California, 1959.

Wright, A. F. (ed.), *The Confucian Persuasion*, Stanford University Press, Stanford, California, 1960.

Yang, Lien-Shêng, *Money and Credit in China; a Short History*, Harvard University Press, Cambridge, Mass., 1952.

Yap, Yong and Cotterell, A., *The Early Civilisation of China*, Putnam, New York, 1975.

Yule, Sir Henry (ed.), *The Book of Ser Marco Polo the Venetian, concerning the Kingdoms and Marvels of the East*, tr. and ed., with notes, by H. Yule, 1st edn 1871, reprinted 1875; 3rd edn, 2 vols., ed. H. Cordier, Murray, London, 1903, reprinted 1921; 3rd edn also issued Scribner, New York, 1929.

Yule, Sir Henry, *Cathay and the Way Thither; being a Collection of Mediaeval Notices of China*, Hakluyt Society, 1st edn 1866, Pubs. 2nd ser., London, 1913–15, revised by H. Cordier in 4 vols.

INDEX